ESSAYS ON MEDICAL FAILURE:
A formal multilevel approach

Rodrick Wallace
Deborah Wallace
Division of Epidemiology
The New York State Psychiatric Institute
rodrick.wallace@gmail.com

October 27, 2011

Preface

Elsewhere we have, as many before us, examined population structures of disease and health in terms of public policies, economic practice and other patterns of resource allocation, and the power relations
between groups, that are all and always embedded in a compelling historical trajectory (Wallace et al., 2009, 2010; Wallace and Fullilove,
2008). This is Western public health, not Western medicine, that remains largely a matter of the therapeutic alliance between practitioner
and patient. Basically, for human populations, good living and working
conditions over sufficient time guarantee long healthy life, with, by and
large, minimal preventive care at the beginning, and palliative care at
the end, of life. Western medical practice is often seen as attempting
to fill the gaps in the absence of such conditions.

Here we turn the analytic machinery developed in those earlier vol-
umes to problems in medicine. By the late 20th Century, Western
medicine – particularly in American circles – had become a quintessen-
tial 'scientific' discipline: One spoke reverently of *the biomedical sciences*, invoking cutting-edge methods from across a broad swath of
biological, physical science, and engineering disciplines. But Western
medicine remains, at base, very much a direct problem-solving enterprise: an individual 'presents' with a spectrum of symptoms, and the
medical practitioner is expected to prescribe treatment to alleviate suffering.

The first chapter introduces generalized communication theory arguments, appropriate, in the sense of Dretske, to interacting cognitive
modules. This permits exploration of how disease states interact with
medical treatment, given an embedding context of structured psychosocial stress. The interpenetrating feedback between treatment and response creates a kind of idiotypic hall of mirrors generating a synergistic pattern of efficacy, treatment failure, adverse reactions, and patient
noncompliance which, from a Rate Distortion perspective, embodies a
distorted image of externally-imposed structured stress. For the US,
accelerating spatial and social spread of such stress enmeshes both dominant and subordinate populations in a linked system of pathogenic
social hierarchy that will express itself, not only in an increasingly unhealthy society, but in the diffusion of therapeutic failure, including,
but not limited to, drug-based treatments.

The second chapter extends a cognitive paradigm for gene expression based on the asymptotic limit theorems of information theory to
the epigenetic epidemiology of mental disorders. In particular, recognizing the fundamental role culture plays in human biology, another
heritage mechanism parallel to, and interacting with, the more familiar genetic and epigenetic systems. This is done via a model through
which culture acts as another tunable epigenetic catalyst that both directs developmental trajectories, and becomes convoluted with individ-

ual ontology, via a mutually-interacting crosstalk mediated by a social interaction that is itself culturally driven. Embedding culture is thus an essential component of the epigenetic regulation of human mental development and its dysfunctions, bringing what is perhaps the central reality of human biology into the center of biological psychiatry. Current US work on gene-environment interactions in psychiatry must be extended to a model of gene-culture-environment interaction to avoid becoming victim of an extreme American individualism that threatens to create paradigms particular to that culture and that are, indeed, peculiar in the context of the world's cultures. The cultural and epigenetic systems of heritage may well provide the 'missing' heritability of complex diseases now under so much intense discussion.

The third chapter examines how mechanistic 'physics' models of protein folding fail to account for the observed spectrum of protein folding and aggregation disorders, suggesting that a more appropriately – and larger scale – biocultural paradigm will be needed for understanding the etiology, prevention, and treatment of these diseases. A nonequilibrium empirical Onsager treatment provides an adaptable statistical model, in the same manner as a regression equation, and produces quasi-equilibrium 'resilience' states representing normal, corrected, eliminated, and pathological states of protein folding. A straightforward generalization to long time scales produces diffusion models for the onset of a large spectrum of protein folding disorders in which epigenetic or life history factors determine the diffusion coefficient or affect the efficiency of chaperone processes.

The fourth chapter examines the devastating 'inverse Moore's Law' of pharmaceutical industry productivity, an exponential increase in the inflation-adjusted cost of bringing a new drug to market over time, rising from \$ 200 million in 1950 to \$1.2 billion by 2010 in a thicket of biological complexity that includes, for humans, a necessary cultural component. An 'information catalysis' perspective on the regulation of biological (and biocultural) processes implies that, while magic bullet interventions are failing to affect complex biological pathologies, multifactorial interventions across scale and level of organization might have synergistic impacts that could evade this 'productivity decline'.

The final chapter discusses the implications of these case histories. Ultimately Western medicine, if taken as a politically palatable replacement for the economic and social reforms that would ensure long, healthy life, has become a difficult engineering discipline, increasingly requiring mathematical tools that would challenge a string theorist. That is, medical practitioners are relentlessly confronted by systematic dynamics of deteriorating population health that cannot be addressed on an individual level with existing or foreseeable technical advances. Nonetheless, the moral injunction to treat the sick remains a powerful motivation, and the methodological innovations we present will likely prove useful.

The book is broadly intended for the collection of physical scientists, pure and applied mathematicians, engineers, computer scientists, and mathematically literate social and biomedical scientists concerned with medicine and public health. The first three chapters are expanded versions of recent peer reviewed articles, (Wallace and Wallace, 2004, 2011; Wallace, 2010), and the fourth is new for this volume. The chapters are largely independent and, except for a few sections that can be skipped at a first reading, generally written at a level common to the physical sciences and applied mathematics.

The implications of this analysis for public policy are significant: Public health is not rocket science, but biomedicine as an ideological, if not practical, substitute for economic and social justice, is a culturally-specific and highly seductive intellectual enterprise, scientifically far more challenging than theoretical physics. In a sense, the Meijins of Western biomedicine can be justly proud of their skills, although the impacts of those skills must greatly diminish as embedding social conditions deteriorate.

References

Wallace, R., and D. Wallace, 2004, Structured psychosocial stress and therapeutic failure, *Journal of Biological Systems*, 12:335-369.

Wallace, R., and D. Wallace, 2011, Cultural epigenetics and the heritability of complex disease, *Transactions on Computational Systems Biology XIII*, LNBI 6575:131-170.

Wallace, R., and M. Fullilove, 2008, *Collective Consciousness and its Discontents: Institutional distributed cognition, racial policy, and public health in the United States*, Springer, New York.

Wallace R., D. Wallace, and R.G. Wallace, 2009, *Farming Human Pathogens: Ecological resilience and evolutionary process*, Springer, New York.

Wallace, R., and D. Wallace, 2010, *Gene Expression and its Discontents: The social production of chronic disease*, Springer, New York.

Wallace, R., 2010, Protein folding disorders: Toward a basic biological paradigm, *Journal of Theoretical Biology*, 267:582-594.

Contents

Chapter 1

Stress and therapeutic failure

1.1 Introduction

Antibiotic therapies are under siege from evolutionary adaptation by pathogens. Malignancies routinely evolve out from under anti-cancer drugs. Antiretroviral therapies against human immunosuppressive virus (HIV), although often able to prolong life, ultimately fail for much the same reason, and multiple-drug-resistant HIV, (MDR-HIV), like MDR-tuberculosis, is an increasing threat. The role of social processes in some of these matters is well understood: for example, public policies of 'planned shrinkage' in New York City (Wallace and Wallace, 1998) created such degrees of social disintegration that many individual courses of drug treatment were repeatedly interrupted, generating MDR strains of TB whose address required an inordinately expensive program of highly labor-intensive directly observed therapy (DOT) estimated to cost $40,000 per patient (Wallace and Wallace, 1997; D. Wallace, 1994).

What is less understood is that the diffusing structured psychosocial stress responsible for the unhealthy societies described by Wilkinson (1996) appears able to affect other therapeutic agents and agencies besides antibiotic drugs. Here we will, in particular, outline how other drug therapies are likely to come under siege, especially in the United States. From a public health perspective this is of limited significance, since salubrious living and working conditions, and egalitarian social relations, rather than effective drugs, are the principal determinants of population health. That being said, spreading decline of drug efficacy and rising rates of adverse reactions and patient noncompliance would nonetheless constitute a significant exacerbation of disease burden.

From the perspective of the economics of the pharmaceutical industry, however, the avalanche of drug failure, side effects, and noncompliance we predict for asthma diabetes, obesity, hypertension, coronary heart disease, hormonal cancers, depression and other mental disorders, is likely to prove devastating. The extraordinary difficulty of bringing new drugs to market will, if we are correct, soon be markedly compounded. Indeed, recent work by Lazarou et al. (1998) suggests that, even at present, adverse drug reactions are already the fourth commonest cause of death in the US.

Adverse drug reactions, to paraphrase Pirmohamed et al. (2002), are typically either consistent with the known pharmacology of the drug, representing an augmentation of its known effects, and are dose-dependent, or else are bizarre responses to idiosyncratic induced hypersensitivity, with highly variable outcomes depending on both the drug and the patient.

The essential role of stress in drug efficacy was recognized some time ago by Downing and Rickels (1982), work rediscovered recently by Haller and colleagues (Haller and Halasz, 2000; Haller, 2001). Downing and Rickels (1982) found that the anxiolytic efficacy of chlordiazepoxide and diazepam was markedly reduced in patients experiencing unfavorable life events during drug treatment, compared with both patients experiencing favorable life events or no major events. As Haller (2001) put it,

> To our knowledge, the impact of this finding [i.e., Downing and Rickels, (1982)] was relatively small, and animal research on the topic was not prompted by it... [nonetheless] these and other similar findings show that drug efficacy is not constant, and [epi]genetic factors (e.g., stress) have a strong modulator effect.

Patient medication noncompliance, to paraphrase the conventional perspective taken by Miura et al. (2001), is already a serious factor limiting the effectiveness of medical treatments. For instance, the classic study of Sackett and Haynes (1976), found that, even under the best of circumstances, over 30 percent of patients skipped their prescribed doses regardless of their disease, prognosis, or symptoms.

Many such problems are related to maintaining long-term therapy in patients with chronic disease such as hypertension. Factors encouraging noncompliance in long-term therapy include the cost of medication, lack of written instructions, nonparticipation of the patient in designing the treatment plan, lack of patient education about disease, side effects, and inconvenient dosing schedules. These factors may enhance the frequency of patient noncompliance as the duration of drug therapy is prolonged.

Langer (1999) takes a view more in concert with current medical anthropology:

> Care providers [have socially constructed] behavior as compliant when it was patterned by their expectations and noncompliant when behavior deviated from their [culturally and socially conditioned] expectations... [Their] [l]ack of awareness of cultural issues increases [patient-provider] social distance, breaks down communications, and precipitates misconceptions between minority patients and their health care providers. Therefore, opportunities for patient dissatisfaction and noncompliance increase...
>
> Brock and Salinsky (1993) use the term 'therapeutic alliance' to denote a process in which the health care provider communicates an assessment of the patient's problem and coordinates a practical management plan that is conducive to patient compliance. This assessment must consider and integrate information about all the systems in which the person exists: biological, psychological, informal/formal social support system, and cultural... Elevating the patient's status within the therapeutic alliance increases the likelihood of... participation, that is, enhanced compliance.

Here we will suggest that stress itself is not undifferentiated, like pressure under water, but often has a complicated grammar and syntax that can, in a sense, carve a distorted image of that structure onto basic human physiological and psychological structures and their responses to medical intervention, including but not limited to, pharmaceuticals.

We will, quite formally, synthesize these matters at and across physiological, psychological, and psychosocial levels of organization, particularly considering the impact of structured stress on the therapeutic alliance.

Essential clues are to be found in observed patterns of population-level disparity in drug efficacy, adverse reactions, and compliance across ethnic groups, once we have discounted contemporary pharmacogenetic ideology.

According to Burroughs et al. (2002), pharmacogenetic research over the past few decades has uncovered significant differences among racial and ethnic groups in the metabolism, clinical effectiveness, and side-effect profiles of many clinically important drugs, a matter which, they claim, can be attributed to "...genetic factors that underlie varying responses to medicines observed among different genetic and racial groups".

An unsigned editorial commentary in *Nature:genetics* (vol. 29, no.3, 239, 2001) is somewhat more guarded, but reaches a similar conclusion, focused, however, at the individual rather than the group locus-of-control:

> Adverse drug reactions or failure to respond to certain drugs can be influenced by polymorphisms in genes responsible for metabolizing drugs. [While] [t]he frequency of such polymorphisms has been found to vary between populations of common ancestry... a number of studies... have shown that average differences in the genetic determinants of drug response... [between] groups... are relatively small and there is considerable overlap between groups. *Once all the genes that contribute to drug response are identified, doctors will be able to prescribe drugs based on patients' [individual] genotypes... this is the promise of the Human Genome Project.* (emphasis added)

Depending on mainstream perspective, then, it is 'genetic polymorphisms' at either the group or individual level that are responsible for most of the variance in patient response to drug therapy.

Currently one of the most discussed examples is the finding that, while 49 percent of white male patients benefited from an angiotensis-converting enzyme inhibitor for heart failure, only 14 percent of African-American males did so (Exner et al., 2001). The point of debate within the biomedical mainstream is not whether this result is a actually attributable to genetic polymorphism or not, but, rather, appears limited to questions regarding the level of organization at which such polymorphism should be investigated, i.e., the racial/ethnic group or the individual.

The reader may have noted the scientific mystery implicit in these results: since the distribution of polymorphisms related to drug efficacy appears to greatly overlap between 'racial' subgroups precisely as the *Nature:genetics* editorial commentary states, the three-fold difference in such efficacy between white and African-American males cannot be entirely, or indeed, even significantly, explained by those polymorphisms.

What is really going on?

Here we raise a larger and far more fundamental locus-of-control question about therapeutic efficacy, compliance, and adverse reactions to therapy, including, but not limited to, drugs. That locus of control is the embedding pattern of structured psychosocial stress exemplified by figure 1.1, a redisplay of material from Singh-Manoux (2003). It shows, for men and women separately, self-reported health as a function of self-reported status rank, where 1 is high and 10 low rank, among some 7000 male and 3400 female London-based office staff, aged 35-55 working in 20 Civil Service departments during the late 1990's. Self-reported health is a highly significant predictor of future morbidity and mortality. The results for both sexes are virtually indistinguishable in what is a kind of toxicological dose-response curve, showing physiological response against a 'dosage' of hierarchy that may include both measures of highly structured stress and real availability of resources (Link and Phelan, 2000). Note that low status staff approach the critical 'ED-50' stage at which half the population shows impact from the dosage.

Figure 1.1 and its generalizations have been the subject of a long scientific debate regarding the way in which the health of those on the lower left hand side are linked to those on the upper right through various mechanisms. Wilkinson, (1996), for example, describes how the rich of nations with lesser disparities between rich and poor are markedly healthier than the rich in nations – like the US – having greater disparities.

In precisely this regard, figure 1.2 shows the Black vs. White diabetes death rates (per 100,000) covering the period 1979-97 for the US, and figure 1.3 the corresponding hypertension death rates over the same time span. 1997 is, in both cases, the point at the upper right. While the Black rates are elevated compared to the White in both cases, the relentless increases are highly correlated indeed, suggesting a single underlying cause which enmeshes both dominant and subordinate populations.

Figure 1.4 suggests etiology. It shows the rapidly increasing percent of total US income accounted for by the highest five percent of earners as a function of the integral of the number of manufacturing jobs lost since 1980. This latter is an index of permanently dispersed social, economic, and political capital for vast sectors of the population. While US deindustrialization has most profoundly affected minority urban neighborhoods (e.g., Massey, 1990), it has also devastated many white working-class communities (e.g., Pappas, 1989). See Wallace et al. (1999) or Wallace and Wallace (2010) for a more complete discussion.

The general inference is that deindustrialization and its associated phenomena have exacerbated a pathogenic social hierarchy enmeshing both dominant and subordinate populations into trajectories of serious developmental disorder. Absent large-scale social, economic, and political reforms, this process will certainly continue.

We intend to generalize this argument to the role of structured psychosocial stress in therapeutic efficacy, failure, adverse reactions, and patient noncompliance.

Within the US, the particular analogs to figure 1.1 inevitably involve the aftermath of a slave economy that persisted into the middle 19th Century and has smoothly evolved into the current, and newly-resurgent, system of American Apartheid (e.g., Massey and Denton, 1993). The latter is inextricably convoluted with a market economy which, in the aftermath of deindustrialization and the decline of unions,

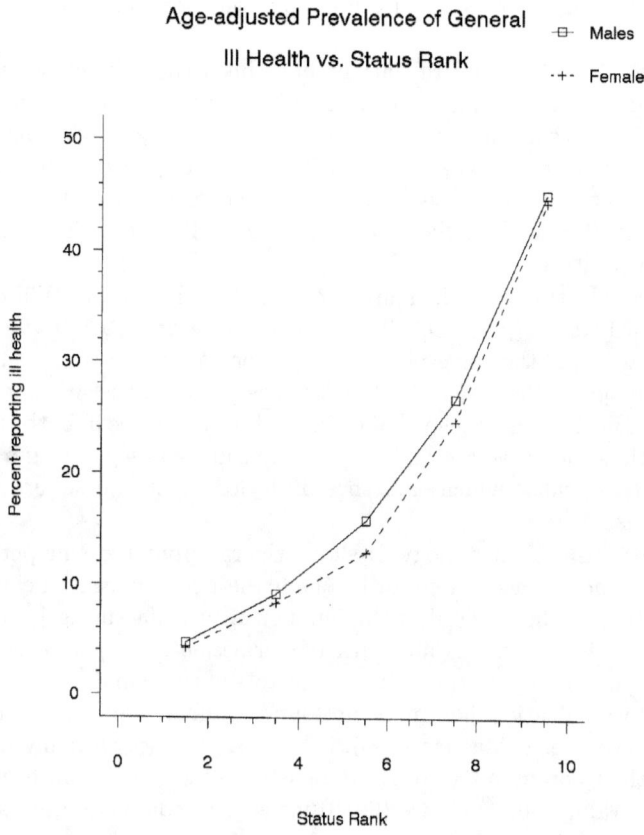

Figure 1.1: Redisplay of data from Singh-Manoux et al. (2003). Sex-specific dose-response curves of age-adjusted prevalence of self-reported ill-health vs. self-reported status rank, Whitehall II cohort, 1997 and 1999. 1 is high and 10 is low status. The curves for male and female are virtually identical, and the upper point is very near the 'EC-50' level in this population i.e., the 'effective concentration' at which fifty percent of subjects show physiological response. Self-reported health is a highly significant predictor of later morbidity and mortality.

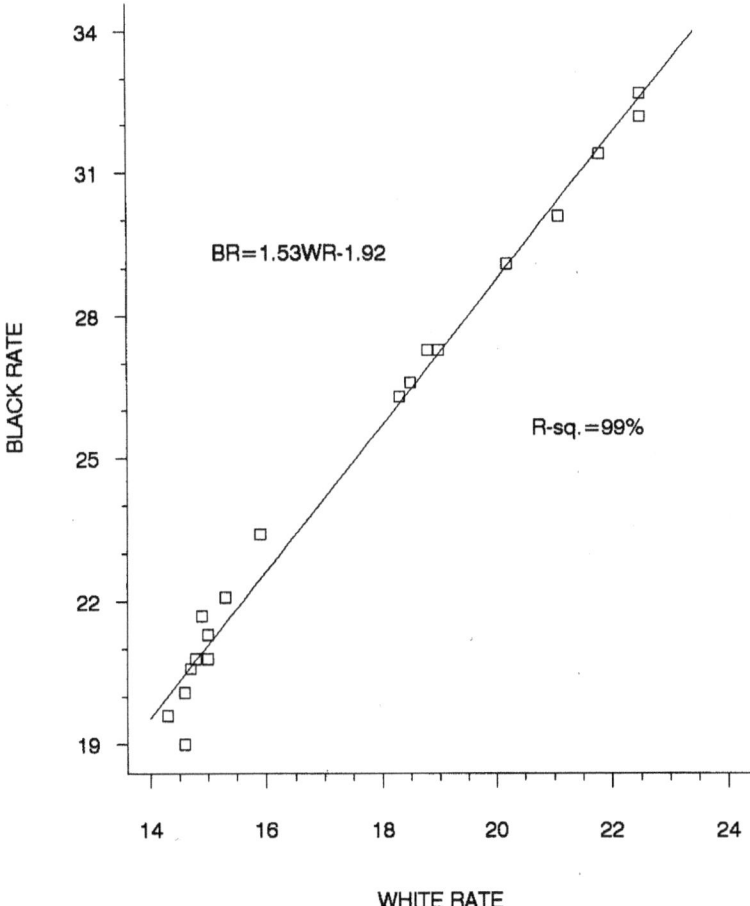

Figure 1.2: US Black vs. White diabetes death rates (per 100,000), 1979-97. While the Black rates are uniformly higher than the White, the coupling between them is very strong indeed, suggesting that, in the words of one researcher, "concentration is not containment" for chronic as well as for infectious diseases like AIDS and tuberculosis.

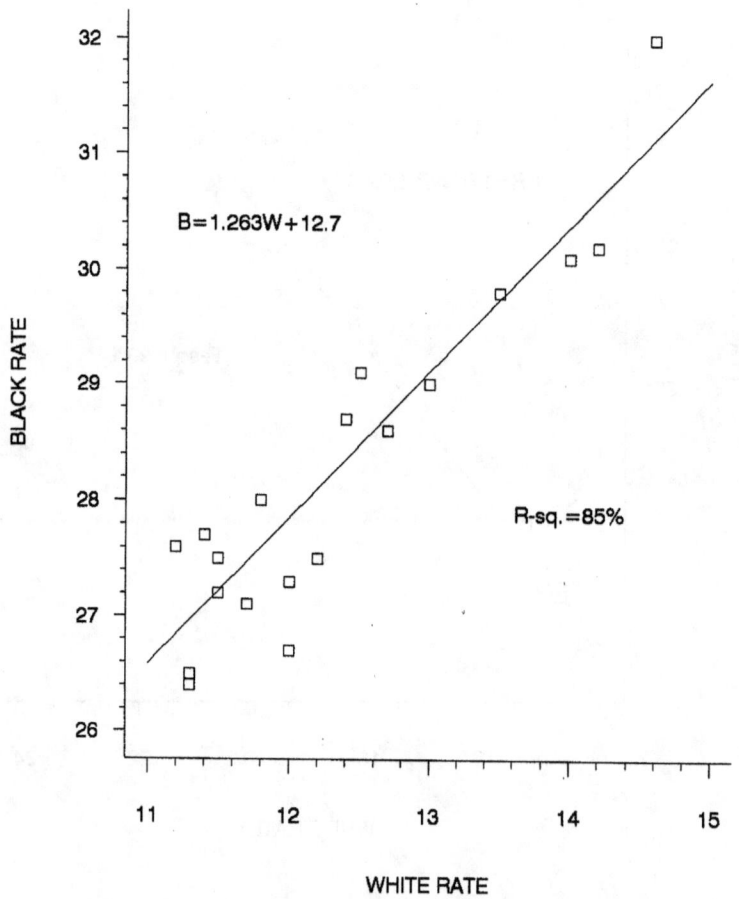

Figure 1.3: Same as figure 1.2 for US hypertension death rates.

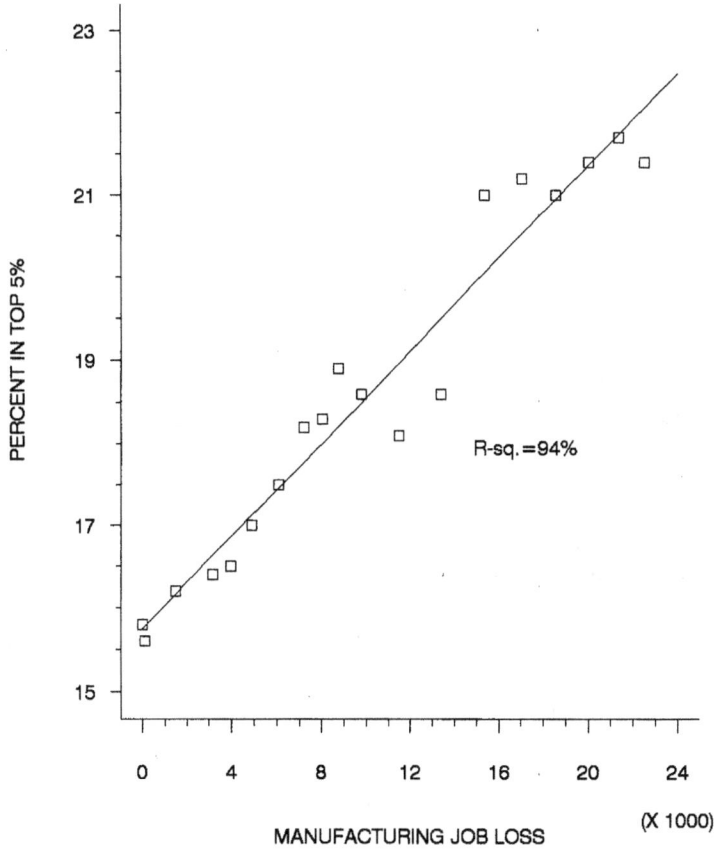

Figure 1.4: Percent of total US income accounted for by the highest five percent of the population vs. integral of manufacturing job loss since 1980. Manufacturing job loss represents the permanent, cumulative, dispersal of social, political, and economic capital for large sectors of the US population. See Wallace and Wallace, 2010 for a more complete discussion.

gives few rights, little stability, and sharply declining real resources for the vast majority of its participants. In addition, deliberate policies of ethnic cleansing ranging from urban renewal to planned shrinkage have left virtually all urban African-American neighborhoods looking like Dresden after the firebombing (e.g., Wallace and Wallace, 1997, 1998), with the inevitable individual and community-scale horrors consequent on such acts.

It is little wonder that African-American males do not respond well to certain classes of drugs. Given the implications of figures 1.2-1.4, it seems unlikely such failure will be restricted to them, however.

We will use a mathematical modeling strategy to investigate more precisely how structured psychosocial stress, like that of figure 1.1, might insinuate itself into physiological and other response to medical therapy, including but not limited to drug efficacy and adverse drug reactions.

The results of this exercise suggest that questions regarding cross sectional and longitudinal structured stress at the individual level might, in the vast majority of cases, provide more useful information in designing and monitoring medical interventions, including drug therapies, than individualized genetic profiles. This conclusion is, however, modulated by the inescapably irreducible consequences of such stress at the population level, a matter to which we will repeatedly return, and that is profoundly implicated in our predictions of spreading drug failure.

We begin by examining mind/body interaction as a composite structure of within-individual 'cognitive modules' enmeshed with an immediate 'sociocultural network', all embedded in a larger context of structured psychosocial stress. We thus implicitly embrace 'comorbidity' between chronic mental and physical disorder, and examine therapeutic intervention in such a structure.

This is no small matter, and leads to use of cutting-edge methods. Indeed, further comment on our methodology is appropriate:

Here we adapt recent advances in understanding punctuated equilibrium in evolutionary process (e.g., Wallace, 2002b; Wallace and Wallace, 1998, 1999) to the question of how embedding structured psychosocial stress affects the interaction of mind, body, pathology, and medical treatment. We specifically seek to determine how the synergism of stress, cognitive submodules, and therapeutic intervention, might be constrained by certain of the asymptotic limit theorems of probability.

We know that, regardless of the probability distribution of a particular stochastic variate, the Central Limit Theorem ensures that long sums of independent realizations of that variate will follow a Normal distribution. Analogous constraints exist on the behavior of information sources – both independent and interacting – and these are described by the limit theorems of information theory. Imposition of phase transition formalism from statistical physics, in the spirit of the Large Deviations Program of applied probability, permits concise and unified description of evolutionary and cognitive 'learning plateaus' which, in the evolutionary case, are interpreted as evolutionary punctuation (e.g., Wallace, 2002a, b). This approach provides a 'natural' means of exploring punctuated processes in the effects of structured stress on mind-body interaction in the context of therapeutic intervention.

The model, as in the relation of the Central Limit Theorem to parametric statistical inference, is almost independent of the detailed structure of the interacting information sources inevitably associated with cognitive process, important as such structure may be in other contexts. This finesses some of the profound ambiguities associated with 'dynamic systems theory' and 'deterministic chaos' treatments in which the existence of 'dynamic attractors' depends on very specific kinds of differential equation models akin to those used to describe ecological population dynamics, chemical processes, or physical systems of weights-on-springs. Cognitive phenomena are neither well-stirred Erlenmeyer flasks of reacting agents, nor distorted mechanical clocks, and the application of nonlinear dynamic systems theory to cognition will likely be found to involve little more than hopeful metaphor. Indeed, much of contemporary nonlinear dynamics can be subsumed within our formalism through symbolic dynamics discretization techniques (Beck and Schlogl, 1995).

In contrast, it seems actually possible to uncover the grammar and syntax of structured psychosocial stress and the function of cognitive submodules, and to express their relations in terms of empirically observed regression models relating measurable biomarkers, behaviors, beliefs, feelings, and so on. Our analysis will focus on the eigenstructure of those models, constrained by the ergodic and other properties of information sources.

Clearly our approach takes much from parametric statistics, and, while idiosyncratic 'nonparametric' models may be required in special cases, we may well capture the essence of the most common relevant phenomena.

This being said, some general remarks regarding the proper use of mathematical models in the conduct of scientific inference are nonetheless appropriate. Much of our reasoning will be based on a fairly elaborate mathematical model of cognitive process. Mathematical models of complicated physiological, social, and other ecosystem phenomena are notorious for their unreliability, instability, and oversimplification. The mathematical ecologist E.C. Pielou (1977, p. 106) puts the matter thus:

...[Mathematical models] are easy to devise; even though

the assumptions of which they are constructed may be hard to justify, the magic phrase 'let us assume that...' overrides objections temporarily. One is then confronted with a much harder task: How is such a model to be tested? The correspondence between a model's predictions and observed events is sometimes gratifyingly close but this cannot be taken to imply the model's simplifying assumptions are reasonable in the sense that neglected complications are indeed negligible in their effects...

In my opinion the usefulness of models is great... [however] it consists *not in answering questions but in raising them*. Models can be used to inspire new field investigations and these are the only source of new knowledge as opposed to new speculation.

In this spirit we will conclude the paper with a number of model-based recommendations for further empirical studies. These are, perhaps, the central product of our enterprise.

1.2   Cognitive modules of human biology

1.2.1   Immune function

Atlan and Cohen (1998) have proposed an information-theoretic cognitive model of immune function and process, a paradigm incorporating cognitive pattern recognition-and-response behaviors analogous to those of the central nervous system. This work follows in a very long tradition of speculation on the cognitive properties of the immune system (e.g., Tauber, 1998; Podolsky and Tauber, 1998; Grossman, 1989, 1992, 1993a, b, 2000).

From the Atlan/Cohen perspective, the meaning of an antigen can be reduced to the type of response the antigen generates. That is, the meaning of an antigen is functionally defined by the response of the immune system. The meaning of an antigen to the system is discernible in the type of immune response produced, not merely whether or not the antigen is perceived by the receptor repertoire. Because the meaning is defined by the type of response there is indeed a response repertoire and not only a receptor repertoire.

To account for immune interpretation Cohen (1992, 2000) has reformulated the cognitive paradigm for the immune system. The immune system can respond to a given antigen in various ways, it has 'options'. Thus the particular response we observe is the outcome of internal processes of weighing and integrating information about the antigen.

In contrast to Burnet's view of the immune response as a simple reflex, it is seen to exercise cognition by the interpolation of a level of information processing between the antigen stimulus and the immune response. A cognitive immune system organizes the information borne by the antigen stimulus within a given context and creates a format suitable for internal processing; the antigen and its context are transcribed internally into the 'chemical language' of the immune system.

The cognitive paradigm suggests a language metaphor to describe immune communication by a string of chemical signals. This metaphor is apt because the human and immune languages can be seen to manifest several similarities such as syntax and abstraction. Syntax, for example, enhances both linguistic and immune meaning.

Although individual words and even letters can have their own meanings, an unconnected subject or an unconnected predicate will tend to mean less than does the sentence generated by their connection.

The immune system creates a 'language' by linking two ontogenetically different classes of molecules in a syntactical fashion. One class of molecules are the T and B cell receptors for antigens. These molecules are not inherited, but are somatically generated in each individual. The other class of molecules responsible for internal information processing is encoded in the individual's germline.

Meaning, the chosen type of immune response, is the outcome of the concrete connection between the antigen subject and the germline predicate signals.

The transcription of the antigens into processed peptides embedded in a context of germline ancillary signals constitutes the functional language of the immune system. Despite the logic of clonal selection, the immune system does not respond to antigens as they are, but to abstractions of antigens-in-context.

1.2.2 Tumor control

We propose that the next cognitive submodule after the immune system is a tumor control mechanism which may include immune surveillance, but clearly transcends it. Nunney (1999) has explored cancer occurrence as a function of animal size, suggesting that in larger animals, whose lifespan grows as about the 4/10 power of their cell count, prevention of cancer in rapidly proliferating tissues becomes more difficult in proportion to size. Cancer control requires the development of additional mechanisms and systems to address tumorigenesis as body size increases – a synergistic effect of cell number and organism longevity. Nunney concludes that this pattern may represent a real barrier to the evolution of large, long-lived animals and predicts that those that do evolve have recruited additional controls over those of smaller animals to prevent cancer.

In particular, different tissues may have evolved markedly different tumor control strategies. All of these, however, are likely to be energetically expensive, permeated with different complex signaling strategies, and subject to a multiplicity of reactions to signals, including those related to psychosocial stress. Forlenza and Baum (2000) explore the effects of stress on the full spectrum of tumor control, ranging from DNA damage and control, to apoptosis, immune surveillance, and mutation rate. Elsewhere (R. Wallace et al., 2003) we argue that this elaborate tumor control strategy, at least in large animals, must be at least as cognitive as the immune system itself, which is one of its components: some comparison must be made with an internal picture of a healthy cell, and a choice made as to response: none, attempt DNA repair, trigger programmed cell death, engage in full-blown immune attack. This is, from the Atlan/Cohen perspective, the essence of cognition.

1.2.3   The HPA axis

The hypothalamic-pituitary-adrenal (HPA) axis, the flight-or-fight system, is clearly cognitive in the Atlan/Cohen sense. Upon recognition of a new perturbation in the surrounding environment, memory and brain or emotional cognition evaluate and choose from several possible responses: no action needed, flight, fight, helplessness (i.e., flight or fight needed, but not possible). Upon appropriate conditioning, the HPA axis is able to accelerate the decision process, much as the immune system has a more efficient response to second pathogenic challenge once the initial infection has become encoded in immune memory. Certainly hyperreactivity in the context of post-traumatic stress disorder (PTSD) is a well known example. Chronic HPA axis activation is deeply implicated in visceral obesity leading to diabetes and heart disease, via the leptin/cortisol diurnal cycle (e.g., Bjorntorp, 2001).

1.2.4   Blood pressure regulation

Rau and Elbert (2001) review much of the literature on blood pressure regulation, particularly the interaction between baroreceptor activation and central nervous function. We paraphrase something of their analysis. The essential point, of course, is that unregulated blood pressure would be quickly fatal in any animal with a circulatory system, a matter as physiologically fundamental as tumor control. Much work over the years has elucidated some of the mechanisms involved: increase in arterial blood pressure stimulates the arterial baroreceptors which in turn elicit the baroreceptor reflex, causing a reduction in cardiac output and in peripheral resistance, returning pressure to its original level. The reflex, however, is not actually this simple: it may be inhibited through peripheral processes, for example under conditions of high metabolic demand. In addition, higher brain structures modulate this reflex arc, for instance when threat is detected and fight or flight responses are being prepared. This suggests, then, that blood pressure control cannot be a simple reflex, but is, rather, a broad and actively cognitive modular system which compares a set of incoming signals with an internal reference configuration, and then chooses an appropriate physiological level of blood pressure from a large repertory of possible levels, i.e., a cognitive process in the Atlan/Cohen sense. The baroreceptors and the baroreceptor reflex are, from this perspective, only one set of a complex array of components making up a larger and more comprehensive cognitive blood pressure regulatory module.

1.2.5 Emotion

Thayer and Lane (2000) summarize the case for what can be described as a cognitive emotional process. Emotions, in their view, are an integrative index of individual adjustment to changing environmental demands, an organismal response to an environmental event that allows rapid mobilization of multiple subsystems. Emotions are the moment-to-moment output of a continuous sequence of behavior, organized around biologically important functions. These 'lawful' sequences have been termed behavioral systems by Timberlake (1994).

Emotions are self-regulatory responses that allow the efficient coordination of the organism for goal-directed behavior. Specific emotions imply specific eliciting stimuli, specific action tendencies including selective attention to relevant stimuli, and specific reinforcers. When the system works properly, it allows for flexible adaptation of the organism to changing environmental demands, so that an emotional response represents a *selection* of an appropriate response and the inhibition of other less appropriate responses from a more or less broad behavioral repertoire of possible responses. Such choice, we will show, leads directly to something closely analogous to the Atlan and Cohen language metaphor.

Damasio (1998) concludes that emotion is the most complex expression of homeostatic regulatory systems. The results of emotion serve the purpose of survival even in nonminded organisms, operating along dimensions of approach or aversion, of appetition or withdrawal. Emotions protect the subject organism by avoiding predators or scaring them away, or by leading the organism to food and sex. Emotions often operate as a basic mechanism for making decisions without the labors of reason, that is, without resorting to deliberated considerations of facts, options, outcomes, and rules of logic. In humans learning can pair emotion with facts which describe the premises of a situation, the option taken relative to solving the problems inherent in a situation, and perhaps most importantly, the outcomes of choosing a certain option, both immediately and in the future. The pairing of emotion and fact remains in memory in such a way that when the facts are considered in deliberate reasoning when a similar situation is revisited, the paired emotion or some aspect of it can be reactivated. The recall, according to Damasio, allows emotion to exert its pairwise qualification effect, either as a conscious signal or as nonconscious bias, or both, In both types of action the emotions and the machinery underlying them play an important regulatory role in the life of the organism. This higher order role for emotion is still related to the needs of survival, albeit less apparently.

Thayer and Friedman (2002) argue, from a dynamic systems perspective, that failure of what they term inhibitory processes which, among other things, direct emotional responses to environmental signals, is an important aspect of psychological and other disorder. Sensitization and inhibition, they claim, sculpt the behavior of an organism to meet changing environmental demands. When these inhibitory processes are dysfunctional – choice fails – pathology appears at numerous levels of system function, from the cellular to the cognitive.

Thayer and Lane (2000) also take a dynamic systems perspective on emotion and behavioral subsystems which, in the service of goal-directed behavior and in the context of a behavioral system, they see organized into coordinated assemblages that can be described by a small number of control parameters, like the factors of factor analysis, revealing the latent structure among a set of questionnaire items thereby reducing or mapping the higher dimensional item space into a lower dimensional factor space. In their view, emotions may represent preferred configurations in a larger 'state-space' of a possible behavioral repertoire of the organism. From their perspective, disorders of affect represent a condition in which the individual is unable to select the appropriate response, or to inhibit the inappropriate response, so that the response selection mechanism is somehow corrupted.

Gilbert (2001) suggests that a canonical form of such 'corruption' is the excitation of modes that, in other circumstances, represent normal evolutionary adaptations, a matter to which we will return at some length below.

1.2.6   Conscious thought

Although the Cartesian dichotomy between rational thought and emotion may be increasingly suspect, nonetheless humans, like many other animals, do indeed conduct individual rational cognitive decision-making as most of us would commonly understand it. Various forms of dementia involve characteristic patterns of degradation in that ability.

Dehaene and Nacache (2001) describe the 'global neuronal workspace' model of conscious rational thought, and our own extension of that model is available elsewhere (Wallace, 2005a, b). Matters regarding mental disorders will be addressed more directly in the next chapter.

1.2.7 Gene expression

Wallace and Wallace (2008, 2009) introduce a cognitive paradigm for gene expression in which developed phenotype is seen as a chosen response to inherited epigenetic and environmental signals during development. Again, this will be the subject of more detailed analysis in the following chapter.

1.2.8 Protein folding

Although amino acid sequence indeed 'predicts' the canonical normal form of a folded protein (e.g., Anfinsen, 1973), the existence of ribosomes, the endoplasmic reticulum, shock protein chaperones, and related mechanisms suggests that there is an elaborate cellular cognitive process that folds, repairs, or eliminates, proteins, according to a cognitive process. This will be the central topic of the third chapter.

1.2.9 Sociocultural network

Humans, however, are particularly noted for a hypersociality which inevitably enmeshes us all in group processes of decision, i.e., collective cognitive behavior within a social network, tinged by an embedding shared culture. For humans, culture is truly fundamental. Durham (1991) argues that genes and culture are two distinct but interacting systems of inheritance within human populations. Information of both kinds has influence, actual or potential, over behaviors, which creates a real and unambiguous symmetry between genes and phenotypes on the one hand, and culture and phenotypes, on the other. Genes and culture are best represented as two parallel lines or tracks of hereditary influence on phenotypes.

Much of hominid evolution can be characterized as an interweaving of genetic and cultural systems. Genes came to encode for increasing hypersociality, learning, and language skills. The most successful populations displayed increasingly complex structures that better aided in buffering the local environment (Bonner, 1980).

Successful human populations seem to have a core of tool usage, sophisticated language, oral tradition, mythology, music, magic, medicine, and religion, and decision making skills focused on relatively small family/extended family social network groupings. More complex social structures are built on the periphery of this basic object (Richerson and Boyd, 1995). The human species' very identity may rest on its unique evolved capacities for social mediation and cultural transmission. These are particularly expressed through the cognitive decision making of small groups facing changing patterns of threat and opportunity, processes in which we are all embedded and all participate.

1.3   Cognition as language

Atlan and Cohen (1998) and Cohen (2000), following a long tradition in the study of immune cognition (e.g., Grossman, 1989; Tauber, 1998), argue that the essence of cognitive function involves comparison of a perceived signal with an internal, learned picture of the world, and then, upon that comparison, the choice of a response from a much larger repertoire of possible responses. Following the approach of Wallace (2000, 2002a), we make a very general model of that process.

Cognitive pattern recognition-and-response, as we characterize it, proceeds by convoluting an incoming external sensory incoming signal with an internal ongoing activity – the learned picture of the world – and triggering an appropriate action based on a decision that the pattern of sensory activity requires a response. We will, fulfilling Atlan and Cohen's (1998) criterion of meaning-from-response, define a language's contextual meaning entirely in terms of system output, leaving out, for the moment, the question of how such a pattern recognition system is trained, a matter for Rate Distortion theory.

The abstract model will be illustrated by two neural network examples.

A pattern of sensory input is mixed in some unspecified but systematic manner with internal 'ongoing' activity to create a path of convoluted signal $x = (a_0, a_1, ..., a_n, ...)$. This path is fed into a highly nonlinear, but otherwise similarly unspecified, decision oscillator which generates an output $h(x)$ that is an element of one of two (presumably) disjoint sets B_0 and B_1 of possible system responses. We take

$$B_0 \equiv \{b_0, ..., b_k\},$$

$$B_1 \equiv \{b_{k+1}, ..., b_m\}.$$

Thus we permit a graded response, supposing that if

$$h(x) \in B_0$$

the pattern is not recognized, and if

$$h(x) \in B_1$$

the pattern is recognized and some action $b_j, k+1 \leq j \leq m$ takes place.

We are interested in paths x which trigger pattern recognition-and-response exactly once. That is, given a fixed initial state a_0, such that $h(a_0) \in B_0$, we examine all possible subsequent paths x beginning with a_0 and leading exactly once to the event $h(x) \in B_1$. Thus $h(a_0, ..., a_j) \in B_0$ for all $j < m$, but $h(a_0, ..., a_m) \in B_1$.

For each positive integer n let $N(n)$ be the number of paths of length n which begin with some particular a_0 having $h(a_0) \in B_0$ and lead to the condition $h(x) \in B_1$. We shall call such paths 'meaningful' and assume $N(n)$ to be considerably less than the number of all possible paths of length n – pattern recognition-and-response is comparatively rare. We further assume that the finite limit

$$H \equiv \lim_{n \to \infty} \frac{\log[N(n)]}{n}$$

both exists and is independent of the path x. We will – not surprisingly – call such a pattern recognition-and-response cognitive process *ergodic*. Not all such processes are likely to be ergodic, implying that H, if it exists, is path dependent, although extension to 'nearly' ergodic processes is straightforward.

Invoking Shannon, we may thus define an ergodic information source $\mathbf{X}$ associated with stochastic variates X_j having joint and conditional probabilities $P(a_0, ..., a_n)$ and $P(a_n|a_0, ..., a_{n-1})$ such that appropriate joint and conditional Shannon uncertainties may be defined which satisfy the relations

$$H[\mathbf{X}] = \lim_{n \to \infty} \frac{\log[N(n)]}{n} =$$

$$\lim_{n \to \infty} H(X_n|X_0, ..., X_{n-1}) =$$

$$\lim_{n \to \infty} \frac{H(X_0, ..., X_n)}{n}.$$

(1.1)

The Shannon uncertainties $H(...)$ are defined in terms of cross-sectional sums of the form $-\sum_k P_k \log[P_k]$, where the P_k constitute a probability distribution. See Ash (1990) or Cover and Thomas (1991) for details.

We say this information source is *dual* to the ergodic cognitive process.

Again, for non-ergodic sources, a limit $\lim_{n\to\infty} H$ may be defined for each path, but it will not necessarily given by the simple cross-sectional law-of-large numbers analogs above. For 'nearly' ergodic systems one might perhaps use something of the form (Wallace and Fullilove, 2007)

$$H(x + \delta x) \approx H(x) + \delta x dH/dx.$$

Different language-analogs will, of course, be defined by different divisions of the total universe of possible responses into different pairs of sets B_0 and B_1, or by requiring more than one response in B_1 along a path. However, like the use of different distortion measures in the Rate Distortion Theorem (e.g. Cover and Thomas, 1991), it seems obvious that the underlying dynamics will all be qualitatively similar.

Similar but not identical, and herein lies the first of two essential matters: dividing the full set of possible responses into sets B_0 and B_1 may itself require higher order cognitive decisions by another module or modules, suggesting the necessity of 'choice' within a more or less broad set of possible languages-of-thought. This would, in one way, reflect the need of the organism to shift gears according to the different challenges it faces, leading to a model for autocognitive disease when a normally excited state is recurrently (and incorrectly) identified as a member of the 'resting' set B_0. We will return to this approach below as a formal alternative to the 'homunculus' regression model treatment given later.

A second possible source of structure, however, lies at the input rather than the output end of the model: i.e., suppose we classify paths instead of outputs. That is, we define equivalence classes in convolutional 'path space' according to whether a state a_k^M can be connected by a path with some originating state a_M. That is, we, in turn, set each possible state to an a_0, and define other states as formally equivalent to it if they can be reached from that (now variable) $a_0 = a_M$ by a grammatical/syntactical path. That is, a state which can be reached by a legitimate path from a_M is taken as equivalent to it. We can thus divide path space into (ordinarily) disjoint sets of equivalence classes. Each equivalence class defines its own language-of-thought: disjoint cognitive modules, possibly associated with an embedding equivalence class algebra, a groupoid structure. The embedding of 'fast' cognitive processes within larger 'slow' cognitive structures leads to a natural nested groupoid formalism that we will pursue further in subsequent chapters. See the Mathematical Appendix for a summary of standard material on groupoids.

While meaningful paths – creating an inherent grammar and syntax – are defined entirely in terms of system response, as Atlan and Cohen (1998) propose, a critical task is to make these (relatively) disjoint cognitive modules interact, and to examine the effects of that interaction on global properties. Punctuated phase transition effects will emerge in a natural manner.

Before proceeding, however, we give two explicit neural network applications.

First the simple stochastic neuron: A series of inputs $y_i^j, i = 1...m$ from m nearby neurons at time j is convoluted with 'weights' $w_i^j, i = 1...m$, using an inner product

$$a_j = \mathbf{y}^j \cdot \mathbf{w}^j = \sum_{i=1}^{m} y_i^j w_i^j$$

in the context of a 'transfer function' $f(\mathbf{y}^j \cdot \mathbf{w}^j)$ such that the probability of the neuron firing and having a discrete output $z^j = 1$ is $P(z^j = 1) = f(\mathbf{y}^j \cdot \mathbf{w}^j)$. Thus the probability that the neuron does not fire at time j is $1 - f(\mathbf{y}^j \cdot \mathbf{w}^j)$.

In the terminology of this section the m values y_i^j constitute 'sensory activity' and the m weights w_i^j the 'ongoing activity' at time j, with $a_j = \mathbf{y}^j \cdot \mathbf{w}^j$ and $x = a_0, a_1, ...a_n, ...$

A little more work leads to a fairly standard neural network model in which the network is trained by appropriately varying the $\mathbf{w}$ through least squares or other error minimization feedback. This can be shown to, essentially, replicate rate distortion arguments (Cover and Thomas, 1991), as we can use the error definition to define a distortion function $d(y, \hat{y})$ which measures the difference between the training pattern y and the network output $\hat{y}$ as a function of, for example, the inverse number of training cycles, K. As discussed in some detail elsewhere (Wallace, 2002), learning plateau behavior follows as a phase transition on the parameter K in the mutual information $I(Y, \hat{Y})$.

Park et al. (2000) treat the stochastic neural network in terms of a space of related probability density functions $[p(\mathbf{x}, \mathbf{y}; \mathbf{w})|\mathbf{w} \in \mathcal{R}^m]$, where $\mathbf{x}$ is the input, $\mathbf{y}$ the output and $\mathbf{w}$ the parameter vector. The goal of learning is to find an optimum $\mathbf{w}^*$ that maximizes the log likelihood function. They define a loss function of learning as

$$L(\mathbf{x}, \mathbf{y}; \mathbf{w}) \equiv -\log p(\mathbf{x}, \mathbf{y}; \mathbf{w}),$$

and one can take as a learning paradigm the gradient relation

$$\mathbf{w}_{t+1} = \mathbf{w}_t - \eta_t \partial L(\mathbf{x}, \mathbf{y}; \mathbf{w})/\partial \mathbf{w},$$

where η_t is a learning rate.

Park et al. (2000) attack this optimization problem by recognizing that the space of $p(\mathbf{x}, \mathbf{y}; \mathbf{w})$ is Riemannian with a metric given by the Fisher information matrix

$$G(\mathbf{w}) = \int \int \partial \log p/\partial \mathbf{w} [\partial \log p/\partial \mathbf{w}]^T p(\mathbf{x}, \mathbf{y}; \mathbf{w}) dy dx$$

where T is the transpose operation. A Fisher-efficient on-line estimator is then obtained by using the 'natural' gradient algorithm

$$\mathbf{w}_{t+1} = \mathbf{w}_t - \eta_t G^{-1} \partial L(\mathbf{x}, \mathbf{y}; \mathbf{w}) / \partial \mathbf{w}.$$

Again, through the synergistic family of probability distributions $p(\mathbf{x}, \mathbf{y}; \mathbf{w})$, this can be viewed as a special case – a 'representation', to use physics jargon – of the general 'convolution argument' given above.

It seems likely that a rate distortion analysis of the interaction between training language and network response language will nonetheless show the ubiquity of learning plateaus, even in this rather elegant special case.

We will eventually parameterize the information source uncertainty of the dual information source with respect to one or more variates, writing, e.g., $H[\mathbf{K}]$, where $\mathbf{K} \equiv (K_1, ..., K_s)$ represents a vector in a parameter space. Let the vector $\mathbf{K}$ follow some path in time, i.e., trace out a generalized line or surface $\mathbf{K}(t)$. We will, following the argument of Wallace (2002b), assume that the probabilities defining H, for the most part, closely track changes in $\mathbf{K}(t)$, so that along a particular 'piece' of a path in parameter space the information source remains as close to memoryless and ergodic as is needed for the mathematics to work. Between pieces, below, we will impose phase transition characterized by a renormalization symmetry, in the sense of Wilson (1971). See Wallace (2005) or Wallace and Wallace (2008) for further details.

We will call such an information source 'adiabatically piecewise stationary ergodic' (APSE).

There are parallels in our development with what Adams (2003) has characterized as 'the informational turn in philosophy', that is, the application of communication theory formalism and concepts to "purposive behavior, learning, pattern recognition, and... the naturalization of mind and meaning", i.e., generalized cognition. One of the first comprehensive attempts was that of Dretske (1981, 1988, 1992, 1993, 1994), whose work Adams describes as follows:

> It is not uncommon to think that information is a commodity generated by things with minds. Let's say that a naturalized account puts matters the other way around, viz. it says that minds are things that come into being by purely natural causal means of exploiting the information in their environments. This is the approach of Dretske as he tried consciously to unite the cognitive sciences around the well-understood mathematical theory of communication...

Dretske himself (1994) writes:

> Communication theory can be interpreted as telling one something important about the conditions that are needed

for the transmission of information as ordinarily understood, about what it takes for the transmission of semantic
information. This has tempted people... to exploit [infor-
mation theory] in semantic and cognitive studies, and thus
in the philosophy of mind.

...Unless there is a statistically reliable channel of communication between [a source and a receiver]... no signal
can carry semantic information... [thus] the channel over
which the [semantic] signal arrives [must satisfy] the appropriate statistical constraints of communication theory.

Here we redirect attention from the informational content or meaning of individual symbols, the province of semantics which so concerned
Dretske, back to the statistical properties of long, internally-structured
paths of symbols emitted by an information source which is 'dual' to a
cognitive process in a particular sense. We will then adapt and mod-
ify a variety of tools from statistical physics to produce dynamically
tunable punctuated or phase transition coupling between interacting
cognitive modules in what we claim is a highly natural manner. As
Dretske so clearly saw, this approach allows scientific inference on the
necessary conditions for cognition, and, we will show, greatly illuminates the global neuronal workspace model of consciousness. It does so
without raising the 18th Century ghosts of noisy, distorted mechanical
clocks inherent to dynamic systems theory.

1.4 Interacting sources

We suppose that the behavior of a cognitive subsystem can be represented by a sequence of states in time, the path $x \equiv x_0, x_1, ...$ Similarly, we assume an external signal of 'structured psychosocial stress'
can also be represented by a path $y \equiv y_0, y_1, ...$ These paths are, however, both very highly structured and, within themselves, are serially
correlated and can, in fact, be represented by 'information sources' $\mathbf{X}$
and $\mathbf{Y}$. We assume the cognitive process and external stressors interact, so that these sequences of states are not independent, but are
jointly serially correlated. We can, then, define a path of sequential
pairs as $z \equiv (x_0, y_0), (x_1, y_1), ...$

The essential content of the Joint Asymptotic Equipartition Theorem, one of the fundamental limit theorems of 20th Century mathematics, is that the set of joint paths z can be partitioned into a relatively
small set of high probability which is termed *jointly typical*, and a much
larger set of vanishingly small probability. Further, according to the
JAEPT, the *splitting criterion* between high and low probability sets
of pairs is the mutual information

$$I(X,Y) = H(X) - H(X|Y) = H(X) + H(Y) - H(X,Y)$$

(1.2)

where $H(X), H(Y), H(X|Y)$ and $H(X,Y)$ are, respectively, the Shannon uncertainties of X and Y, their conditional uncertainty, and their joint uncertainty. See Cover and Thomas (1991) or Ash (1990) for mathematical details. As stated above, the Shannon-McMillan Theorem and its variants permit expression of the various uncertainties in terms of cross sectional sums of terms of the form $-P_k \log[P_k]$ where the P_k are appropriate direct or conditional probabilities. Similar approaches to neural process have been recently adopted by Dimitrov and Miller (2001).

The high probability pairs of paths are, in this formulation, all equiprobable, and if $N(n)$ is the number of jointly typical pairs of length n, then, according to the Shannon-McMillan Theorem and its 'joint' variants,

$$I(X,Y) = \lim_{n\to\infty} \frac{\log[N(n)]}{n}.$$

(1.3)

Generalizing the earlier language-on-a-network models of Wallace and Wallace (1998, 1999), we suppose there is a 'coupling parameter' P representing the degree of linkage between the cognitive system of interest and the structured 'language' of external signals and stressors, and set $K = 1/P$, following the development of those earlier studies. Then we have

$$I[K] = \lim_{n\to\infty} \frac{\log[N(K,n)]}{n}.$$

The essential homology between information theory and statistical mechanics lies in the similarity of this expression with the infinite volume limit of the free energy density. If $Z(K)$ is the statistical mechanics partition function derived from the system's Hamiltonian, then the free energy density is determined by the relation

$$F[K] = \lim_{V \to \infty} \frac{1}{K} \frac{\log[Z(K,V)]}{V} \equiv \frac{\log[\hat{Z}(K,V)]}{V}.$$

(1.4)

F is the free energy density, V the system volume and $K = 1/T$, where T is the system temperature.

We and others argue at some length (e.g., Wallace and Wallace, 1998, 1999; Wallace, 2000; Rojdestvensky and Cottam, 2000; Feynman, 1996) that this is indeed a systematic mathematical homology which, we contend, permits importation of renormalization symmetry into information theory. Imposition of invariance under renormalization on the mutual information splitting criterion $I(X,Y)$ implies the existence of phase transitions analogous to learning plateaus or punctuated evolutionary equilibria in the relations between cognitive mechanism and external perturbation. An extensive mathematical treatment of these ideas is presented elsewhere (Wallace, 2000, 2002a, b; Wallace, 2005; Wallace and Wallace, 2008).

Elaborate developments are possible. From a the more limited perspective of the Rate Distortion Theorem, a selective corollary of the Shannon-McMillan Theorem, we can view the onset of a punctuated interaction between the cognitive mechanism and external stressors as the literal writing of distorted image of those stressors upon cognition:

Suppose that two (adiabatically, piecewise stationary) ergodic information sources $\mathbf{Y}$ and $\mathbf{B}$ begin to interact, to talk to each other, i.e., to influence each other in some way so that it is possible, for example, to look at the output of $\mathbf{B}$ – strings b – and infer something about the behavior of $\mathbf{Y}$ from it – strings y. We suppose it possible to define a retranslation from the B-language into the Y-language through a deterministic code book, and call $\hat{\mathbf{Y}}$ the translated information source, as mirrored by $\mathbf{B}$.

Define some distortion measure comparing paths y to paths $\hat{y}$, $d(y,\hat{y})$ (Cover and Thomas, 1991). We invoke the Rate Distortion Theorem's mutual information $I(Y,\hat{Y})$, which is the splitting criterion between high and low probability pairs of paths. Impose, now, a parameterization by an inverse coupling strength K, and a renormalization symmetry representing the global structure of the system coupling.

Extending the analyses, triplets of sequences, Y_1, Y_2, Z, for which one in particular, here Z, is the 'embedding context' affecting the other two, can also be divided by a splitting criterion into two sets, having high and low probabilities respectively. The probability of a particular triplet of sequences is then determined by the conditional probabilities

$$P(Y_1 = y^1, Y_2 = y^2, Z = z) = \Pi_{j=1}^n p(y_j^1 | z_j) p(y_j^2 | z_j) p(z_j).$$

(1.5)

That is, Y_1 and Y_2 are, in some measure, driven by their interaction with Z.

For large n the number of triplet sequences in the high probability set will be determined by the relation (Cover and Thomas, 1992, p. 387)

$$N(n) \propto \exp[n I(Y_1; Y_2 | Z)],$$

(1.6)

where splitting criterion is given by

$$I(Y_1; Y_2 | Z) \equiv$$

$$H(Z) + H(Y_1 | Z) + H(Y_2 | Z) - H(Y_1, Y_2, Z).$$

We can then examine mixed cognitive/adaptive phase transitions analogous to learning plateaus (Wallace, 2002b) in the splitting criterion $I(Y_1, Y_2 | Z)$. Note that our results are almost exactly parallel to the Eldredge/Gould model of evolutionary punctuated equilibrium (Eldredge, 1985; Gould, 2002).

We can, for the purposes of this work, extend this model to any number of interacting information sources, $Y_1, Y_2, ..., Y_s$ conditional on an external context Z in terms of a splitting criterion defined by

$$I(Y_1; ...; Y_s | Z) = H(Z) + \sum_{j=1}^{s} H(Y_j | Z) - H(Y_1, ..., Y_s, Z),$$

(1.7)

where the conditional Shannon uncertainties $H(Y_j|Z)$ are determined by the appropriate direct and conditional probabilities.

1.5 The generalized retina

Cohen (2000) argues for an 'immunological homunculus' as the immune system's perception of the body as a whole. The particular utility of such a thing, in his view, is that sensing perturbations in a bodily self-image can serve as an early warning sign of pending necessary inflammatory response – expressions of tumorigenesis, acute or chronic infection, parasitization, and the like. Thayer and Lane (2000) argue something analogous for emotional response as a quick internal index of larger patterns of threat or opportunity.

It seems obvious that the collection of interacting cognitive submodules we have explored above must also have a coherent internal self-image of the state of the mind/body and its social relationships. This inferred picture, at the individual level, we term the 'generalized retina' (GR). We shall use the responses of the GR to characterize physiological/mental responses to both illness and to medical interventions, including drugs, used to treat that illness. Illness and treatment may come to reflect one another in a hall of mirrors reminiscent of Jerne's idiotypic network proposed for the dynamics of the immune system.

Let us suppose we cannot measure either stress or cognitive submodule function directly, but can determine the concentrations of hormones, neurotransmitters, certain cytokines, and other biomarkers, or else macroscopic behaviors, beliefs, feelings, or other responses associated with the function of cognitive submodules according to some natural time frame inherent to the system. This would typically be the circadian cycle in both men and women, and the hormonal cycle in premenopausal women. Suppose, in the absence of extraordinary meaningful psychosocial stress, we measure a series of n biomarker concentrations, behavioral characteristics, other indices at time t which we represent as an n-dimensional vector X_t. Suppose we conduct a number of experiments, and create a regression model so that we can, in the absence of perturbation, write, to first order, the markers at time $t + 1$ in terms of that at time t using a matrix equation of the form

$$X_{t+1} \approx \mathbf{R} X_t,$$

(1.8)

where $\mathbf{R}$ is the matrix of regression coefficients, and we have normalized to a zero vector of constant terms.

Suppose we write a GR response to short-term perturbation as

$$X_{t+1} = (\mathbf{R}_0 + \delta\mathbf{R}_{t+1})X_t,$$

where $\delta\mathbf{R}$ represents variation of the generalized cognitive self-image about the basic state $\mathbf{R}_0$.

We impose a (Jordan block) diagonalization in terms of the matrix of (generally nonorthogonal) eigenvectors $\mathbf{Q}_0$ of some 'zero reference state' $\mathbf{R}_0$, obtaining, for an initial condition which is an eigenvector $Y_t \equiv Y_k$ of $\mathbf{R}_0$,

$$Y_{t+1} = (\mathbf{J}_0 + \delta\mathbf{J}_{t+1})Y_k = \lambda_k Y_k + \delta Y_{t+1} =$$

$$\lambda_k Y_k + \sum_{j=1}^{n} a_j Y_j,$$

(1.9)

where $\mathbf{J}_0$ is a (block) diagonal matrix as above, $\delta\mathbf{J}_{t+1} \equiv \mathbf{Q}_0\delta\mathbf{R}_{t+1}\mathbf{Q}_0^{-1}$, and δY_{t+1} *has been expanded in terms of a spectrum of the eigenvectors of* $\mathbf{R}_0$, with

$$|a_j| \ll |\lambda_k|, |a_{j+1}| \ll |a_j|.$$

(1.10)

The essential point is that, provided $\mathbf{R}_0$ has been properly tuned, so that this condition is true, the first few terms in the spectrum of the plieotropic iteration of the eigenstate will contain almost all of the essential information about the perturbation, i.e., most of the variance. We envision this as similar to the detection of color in the optical retina, where three overlapping non-orthogonal 'eigenmodes' of response suffice to characterize a vast array of color sensations. Here, if a concise spectral expansion is possible, a very small number of (typically nonorthogonal) 'generalized cognitive eigenmodes' permit characterization of a vast range of external perturbations, and rate distortion constraints become very manageable indeed. Thus GR responses – the spectrum of excited eigenmodes of $\mathbf{R}_0$, provided it is properly tuned – can be a very accurate and precise gauge of environmental perturbation.

The choice of zero reference state $\mathbf{R}_0$, the 'base state' from which perturbations are measured, is, we claim, a highly nontrivial task, necessitating a specialized apparatus.

This is no small matter. According to current theory, the adapted human mind functions through the action and interaction of distinct mental modules which evolved fairly rapidly to help address special problems of environmental and social selection pressure faced by our Pleistocene ancestors (Barkow et al., 1992). Here we have postulated the necessity of other physiological and social cognitive modules. As is well known in computer engineering, calculation by specialized submodules – numeric processor chips – can be a far more efficient means of solving particular well-defined classes of problems than direct computation by a generalized system. We suggest, then, that our generalized cognition has evolved specialized submodules to speed the address of certain commonly recurring challenges. Nunney (1999) has argued that, as a power law of cell count, specialized subsystems are increasingly required to recognize and redress tumorigenesis, mechanisms ranging from molecular error-correcting codes, to programmed cell death, and finally full-blown immune attack.

We argue that identification of the designated normal state of the GR – generalized cognition's self-image of the body and its social relationships – is difficult, requiring a dedicated cognitive submodule within overall generalized cognition. This is essentially because, for the vast majority of information systems, unlike mechanical systems, there are no restoring springs whose low energy state automatically identifies equilibrium: relatively speaking, all states of the GR are 'high energy' states. That is, active comparison must be made of the state of the GR with some stored internal reference picture, and a decision made about whether to reset to zero, which is a cognitive process. We further speculate that the complexity of such a submodule must also follow something like Nunney's power law with animal size, as the overall generalized cognition and its image of the self, become increasingly complicated with rising number of cells and levels of linked cognition.

Failure of that cognitive submodule can result in identification of an excited state of the GR as normal, triggering the collective patterns of systemic activation which, following the argument of Wallace

(2003g), constitute certain comorbid mental and chronic physical disorders. This would result in a relatively small number of characteristic eigenforms of comorbidity, which would typically become more mixed with increasing disorder.

In sum, since such 'zero mode identification' (ZMI) is a (presumed) cognitive submodule of overall generalized cognition, it involves convoluting incoming 'sensory' with 'ongoing' internal memory data in choosing the zero state, i.e., defining R_0. The dual information source defined by this cognitive process can then interact in a punctuated manner with 'external information sources' according to the Rate Distortion and related arguments above. From a RDT perspective, then, those external information sources literally write a distorted image of themselves onto the ZMI, often in a punctuated manner: (relatively) sudden onset of a developmental trajectory to comorbid mental disorders and chronic physical disease.

Different systems of external signals – including but not limited to structured psychosocial stress – will, presumably, write different characteristic images of themselves onto the ZMI cognitive submodule, i.e., trigger different patterns of comorbid mental/physical disorder.

A brief reformulation in abstract terms may be of interest. Recall that the essential characteristic of cognition in our formalism involves a function h which maps a (convolutional) path $x = a_0, a_1, ..., a_n, ...$ onto a member of one of two disjoint sets, B_0 or B_1. Thus respectively, either (1) $h(x) \in B_0$, implying no action taken, or (2), $h(x) \in B_1$, and some particular response is chosen from a large repertoire of possible responses. We discussed briefly the problem of defining these two disjoint sets, and suggested that some 'higher order cognitive module' might be needed to identify what constituted B_0, the set of 'normal' states. Again, this is because there is no low energy mode for information systems: virtually all states are more or less high energy states, and there is no way to identify a ground state using the physicist's favorite variational or other minimization arguments on energy.

Suppose that higher order cognitive module, which we now recognize as a kind of Zero Mode Identification, interacts with an embedding language of structured psychosocial stress (or other systemic perturbation) and, instantiating a Rate Distortion image of that embedding stress, begins to include one or more members of the set B_1 into the set B_0. Recurrent 'hits' on that aberrant state would be experienced as episodes of highly structured comorbid mind/body pathology.

Empirical tests of this hypothesis, however, quickly lead again into real-world regression models involving the interrelations of measurable biomarkers, beliefs, behaviors, feelings, and so on, requiring formalism much like that used above. The GR can, then, be viewed as a generic heuristic device typifying such regression approaches.

The retina approach is more appropriately characterized as a 'Rate

Distortion Manifold', a local projection that, through overlap, has global structure, much like the tangent planes to a complicated geometric object. Glazebrook and Wallace (2009a, b) provide more detailed, indeed cutting-edge, mathematical treatment.

1.6 Therapeutic efficacy

To reiterate, if $\mathbf{X}$ represents the information source dual to 'zero mode identification' in generalized cognition, and if $\mathbf{Z}$ is the information source characterizing structured psychosocial stress, which constitutes an embedding context, the mutual information between them

$$I(\mathbf{X}; \mathbf{Z}) = H(\mathbf{X}) - H(\mathbf{X}|\mathbf{Z})$$

(1.11)

serves as a splitting criterion for pairs of linked paths of states.

We suppose it possible to parameterize the coupling between these interacting information sources by some inverse coupling parameter, K, writing

$$I(\mathbf{X}; \mathbf{Z}) = I[K],$$

(1.12)

with structured psychosocial stress as the embedding context.

Invocation of the mathematical homology between equations (1.4) and (1.5) permits imposition of renormalization formalism (Wallace, 2000; Wallace et al., 2003a; Wallace and Wallace, 2008) resulting in punctuated phase transition depending on K.

Socioculturally constructed and structured psychosocial stress, in this model having both (generalized) grammar and syntax, can be viewed as entraining the function of zero mode identification when the coupling with stress exceeds a threshold. More than one threshold appears likely, accounting in a sense for the typically staged nature of environmentally caused disorders. These should result in a synergistic – i.e., comorbidly excited – mixed affective, rationally cognitive, psychosocial, and inflammatory or other physical excited state of otherwise normal response, and represent the effect of stress on the linked decision processes of various cognitive functions, in particular through the identification of a false 'zero mode' of the GR. This is a collective, but highly systematic, 'tuning failure' which, in the Rate Distortion sense, represents a literal image of the structure of imposed psychosocial stress written upon the ability of the GR to characterize a normal condition of excitation, causing a mixed excited state of chronically comorbid mental and physical disorder.

In this model different eigenmodes Y_k of the GR regression model characterized by the matrix $\mathbf{R}_0$ can be taken to represent the 'shifting-of-gears' between different 'languages' defining the sets B_0 and B_1. That is, different eigenmodes of the GR would correspond to different required (and possibly mixed) characteristic systemic responses.

If there is a state (or set of states) Y_1 such that $\mathbf{R}_0 Y_1 = Y_1$, then the 'unitary kernel' Y_1 corresponds to the condition 'no response required', the set B_0.

Suppose pathology becomes manifest, i.e.,

$$\mathbf{R}_0 \rightarrow \mathbf{R}_0 + \delta\mathbf{R} \equiv \hat{\mathbf{R}}_0,$$

so that some chronic excited state becomes the new 'unitary kernel', and

$$Y_1 \rightarrow \hat{Y}_1 \neq Y_1$$

$$\hat{\mathbf{R}}_0 \hat{Y}_1 = \hat{Y}_1.$$

This could represent, for example, chronic inflammation, autoimmune response, persistent depression/anxiety or HPA axis activation/burnout, and so on.

We wish to induce a sequence of therapeutic counterperturbations $\delta\mathbf{T}_k$ according to the pattern

$$[\hat{\mathbf{R}}_0 + \delta\mathbf{T}_1]\hat{Y}_1 = Y^1,$$

$$\hat{\mathbf{R}}_1 \equiv \hat{\mathbf{R}}_0 + \delta\mathbf{T}_1,$$

$$[\hat{\mathbf{R}}_1 + \delta\mathbf{T}_2]Y^1 = Y^2$$

...

(1.13)

so that, in some sense,

$$Y^j \to Y_1.$$

(1.14)

That is, the mind/body system, as monitored by the GR, is driven to its original condition.

We may or may not have $\hat{\mathbf{R}}_0 \to \mathbf{R}_0$. That is, actual cure may not be possible, in which case palliation or control is the therapeutic aim.

The essential point is that the pathological state represented by $\hat{\mathbf{R}}_0$ and the sequence of therapeutic interventions $\delta\mathbf{T}_k, k = 1, 2, ...$ are interactive and reflective, depending on the regression of the set of vectors Y^j to the desired state Y_1, much in the same spirit as Jerne's immunological idiotypic hall of mirrors.

The therapeutic problem revolves around minimizing the difference between Y^k and Y_1 over the course of treatment: that difference represents the inextricable convolution of 'treatment failure' with 'adverse reactions' to the course of treatment itself, and 'failure of compliance' attributed through social construction by provider to patient, i.e., failure of the therapeutic alliance.

It should be obvious that the treatment sequence $\delta\mathbf{T}_k$ represents a cognitive path of interventions which has, in turn, a dual information source in the sense we have previously invoked.

Treatment may, then, interact in the usual Rate Distortion manner with patterns of structured psychosocial stress which are, themselves, signals from an embedding information source. Thus treatment failure, adverse reactions, and patient noncompliance will, of necessity, embody a distorted image of structured psychosocial stress.

In sum, characteristic patterns of treatment failure, adverse reactions, and patient noncompliance reflecting collapse of the therapeutic alliance, will occur in virtually all therapeutic interventions according to the manner in which structured psychosocial stress is expressed as an image within the treatment process. This would most likely occur in a highly punctuated manner, depending in a quantitative way on the degree of coupling of the three-fold system of affected individual, patient/provider interaction, and treatment mode, with that stress.

Given that the principal environment of humans is defined by interaction with other humans and with socioeconomic institutions, these
are likely to be very strong effects indeed.

We provide a non-pharmaceutical case history.

1.7 Malaria and the Fulani

Modiano et al. (1996, 1998, 2001a, b) conducted comparative surveys
on three roughly co-resident West African ethnic groups exposed to
the same strains of malaria. The Fulani, Mossi, and Rimaibe live in
the same conditions of hyperendemic transmission in a Sudan savanna
area northeast of Ouagadougou, Burkina Faso. The Mossi and Rimaibe
are Sudanese Negroid populations with a long tradition of sedentary
farming, while the Fulani are nomadic pastoralists, partly settled and
characterized by non-Negroid features of possible Caucasoid origin.

Parasitological, clinical, and immunological investigations showed
consistent interethnic differences in *P. Falciparum* infection rates, malaria
morbidity, and prevalence and levels of antibodies to various *P. Falciparum* antigens. The data point to a remarkably similar response to
malaria in the Mossi and Rimaibe, while the Fulani are clearly less parasitized, less affected by the disease, and more responsive to all antigens
tested. No difference in the use of malaria protective measures was
demonstrated that could account for these findings. Known genetic
factors of resistance to malaria showed markedly *lower* frequencies in
the Fulani (Modiano et al, 2001a, b). The differences in the immune
response were not explained by the entomological observations, which
indicated substantially uniform exposure to infective bites.

In their first study, Modiano et al. (1996) concluded that sociocultural factors are not involved in this disparity, and that the available
data support the existence of unknown genetic factors, possibly related
to humoral immune responses, determining interethnic differences in
the susceptibility to malaria.

In spite of later finding the Fulani in their study region have significantly *reduced* frequencies of the classic malaria-resistance genes
compared to the other ethnic groups, Modiano et al. (2001a, b) again
concluded that their evidence supports the existence in the Fulani of
unknown genetic factor(s) of resistance to malaria.

This vision of strict genetic causality carries consequences, seriously
constraining interpretation of the efficacy of interventions. Modiano
et al. (1998) report results of an experiment in their Burkina Faso
study zone involving the distribution of permethrin-impregnated curtains (PIC) to the three co-resident populations, with markedly different results:

The PIC were distributed in June 1996 and their impact

on malaria infection was evaluated in [the three] groups whose baseline levels of immunity to malaria differed because of their age and ethnic group. Age- and ethnic-dependent efficacy of the PIC was observed. Among Mossi and Rimaibe, the impact (parasite rate reduction after PIC installation with respect to the pre-intervention surveys) was 18.8 % and 18.5 %, respectively. A more than two-fold general impact (42.8 %) was recorded in the Fulani. The impact of the intervention on infection rates appears positively correlated with the levels of anti-malaria immunity...

Modiano et al. (1998) conclude from this experiment that the expected complementary role of a hypothetical vaccine is presaged by these results, which also, in their view, emphasize the importance of the genetic background of the population in the evaluation and application of malaria control strategies.

While we fully agree with the importance of their results for a hypothetical vaccine, much in the spirit of Lewontin (2000), we beg to differ with the ad hoc presumptions of genetic causality, which paper over alternatives involving environment and development consistent with these observations.

Recently a medical anthropologist, Andrew Gordon (2000), published a detailed study of Fulani cultural identity and illness:

Cultural identity – who the Fulani think they are – informs thinking on illnesses they suffer. Conversely, illness, so very prevalent in sub-Saharan Africa, provides Fulani with a consistent reminder of their distinctive condition... How they approach being ill also tells Fulani about themselves. The manner in which Fulani think they are sick expresses their sense of difference from other ethnic groups. Schemas of [individual] illness and of collective identity draw deeply from the same well and web of thoughts... As individuals disclose or conceal illness, as they discuss illness and the problem of others, they reflect standards of Fulani life – being strong of character not necessarily of body, being disciplined, rigorously Moslem, and leaders among lessors... to be in step with others and with cultural norms is to have pride in the self and the foundations of Fulani life.

The Fulani carried the Islamic invasion of Africa into the sub-Sahara, enslaving and deculturing a number of ethnic groups, and replacing the native languages with their own. This is much the way African Americans were enslaved, decultured, and taught English.

As Gordon puts it,

'True Fulani' see themselves as distinguished by their aristocratic descent, religious commitments, and personal qualities that clearly differ from lowland cultivators. Those in the lowland are, historically, Fulani subjects who came to act like and speak Fulani, but they are thought to be without the right genealogical descent. The separation between pastoralists and agriculturists repeats itself in settlements across Africa. The terms vary from place to place in Guinea, the terms are Fulbhe for the nobles and the agriculturalist Bhalebhe or Maatyubhee; in Burkina Faso, Fulbhe and the agricultural Rimaybhe; and in Nigeria, the Red Fulani and the agricultural Black Fulani... The schemas for the Fulani body describe the differences between them and others. These are differences that justify pride in being Fulani and not Bhalebhe, Maatyubhe, Rimaybhe, or Black Fulani. In Guinea, the word 'Bhalebhe' means 'the black one'. The term 'Bhalebhe' carries the same meaning as 'Negro' did for Africans brought to North America. It effaces any tribal identity...

The control a Fulani exercises over the body is an essential feature of 'the Fulani way.' Being out of control is shameful and not at all Fulani-like... To act without restraint is to be what is traditionally thought of as Bhalebhe...

Being afflicted with malaria – and handling it well – is a significant proof of ethnicity. How Fulani handle malaria may be telling. What they lack in physical resistance to disease they make up in persistence. Though sickly, Fulani men only reluctantly give into malaria and forgo work. To give into physical discomfort is not *dimo*. When malaria is severe for a man he is likely not to succumb to bed, but instead to sit outside of his home socializing.

The contrasting Occam's Razor hypothesis to genetic determinism, then, is that the observed significant difference in both malarial parasitization and the efficacy of non-pharmaceutical intervention between the historically-dominant Fulani and co-resident historically-subordinate ethnic groups in the Ouagadougou region of Burkina Faso is largely accounted for by longitudinal and cross-sectional factors of structured psychosocial stress, synergistically intersecting with medical intervention, particularly in view of the *lower* frequencies of classic malaria-resistance genes found in the Fulani.

It is not that the Fulani aren't parasitized, or that the 'Fulani way' prevents disease, but that the population-level burdens of environment are modulated by historical development, and these are profoundly different for former masters and former slaves.

1.8 Therapeutic failure in the US

Structured psychosocial stress can, from our development, write an image of itself onto the success or failure of individual-level therapeutic intervention, drug-related or not. The punctuated nature of individual-level response to the coupling with structured stress should, when averaged across a population, reflect itself in a nonlinear 'dose-response' relation between environmental indices of stress and indices of physiological impact, much as in figure 1.1. Apartheid systems, which structure stress in 'Western' societies, following the models of Fanon (1966), Memmi, (1967, 1969), or Wilkinson (1996), are generally seen as frozen, Manichean, structures entangling dominant and subordinate populations in a synergistically dehumanizing pathogenic relationship adversely affecting both, although the powerful are, as always, relatively healthier than those they dominate.

Within the United States, however, the apartheid system is changing in such a way as to enmesh increasingly larger populations in its outfalls. Our own studies show how the post WW-II 'hollowing out' of major US urban centers through contagious urban decay and other forms of policy-driven 'urban desertification' has pumped social disintegration – carrying with it both contagious disease and behavioral pathology – into the commuting field surrounding central cities. These phenomena, in turn, have likewise diffused down along the US urban hierarchy itself, from larger to smaller metropolitan regions (e.g., D. Wallace and R. Wallace, 1998; R. Wallace and D. Wallace, 1997; D. Wallace, 2001). R. Wallace et al. (1997) explicitly demonstrate, for eight large US urban centers, how spreading social disintegration entrained AIDS, tuberculosis, and violent crime into surrounding suburban counties, the most 'hollowed out' central cities being the most pathogenic.

At present, chronic disease epidemics of asthma, obesity, and diabetes, are spreading across the US, even marching up the social hierarchy. It is our contention that these represent the enmeshment of more and more subpopulations into relations of pathogenic social hierarchy like figure 1.1 (e.g., Wallace et al., 2003c, d).

Figures 1.2-1.4 display data related to that argument.

We predict that, not only will chronic developmental disorders continue to diffuse socially and geographically, but failures of drug efficacy and patient compliance, and adverse drug reactions, which in a literal sense mirror the same enmeshing patterns of structured psychosocial stress creating the disorders, will likewise diffuse between and within both social subgroups and geographic regions in a similar manner. Current ideologies of genetic determinism however, have created significant impediments to empirical studies of this hypothesis. These ideological constraints on science may, in the long run, prove very expensive indeed to the pharmaceutical industry. The example of Lysenkoism in the agriculture of the Soviet Union may well be mirrored by the effects of ideological geneticism in Western medicine.

In sum, slavery, racism, apartheid, ethnic cleansing, and their collective aftermath, in concert with draconian economic inequality compounded by deindustrialization in the aftermath of the Cold War, form the context for patterns of both disease onset and progression, and medical treatment efficacy affecting ethnic minorities in the United States and other subject populations elsewhere in the world. This is a context which is not containable within those populations, but enmeshes dominant subgroups as well (Memmi, 1967, 1969), as exemplified by figures 1.2-1.4.

The result of Exner et al. (2001) regarding a more than three-fold difference in heart failure drug efficacy between white and African-American males in the US does not need invocation of genetic differences for its explanation. Following Lewontin's (2000) lead, we invoke instead the triple helix of genes, environment and development, in both disease and in response to, and compliance with, medical intervention against disease. Given that genetic differences between chimpanzees are generally far greater than genetic differences between humans, the Occam's Razor explanation for differences in response to medical intervention would seem to lie primarily in matters of environment and development.

Independent of the mathematical modeling exercise which has led us to this conclusion – a conclusion others have reached without recourse to mathematics, (Braun, 2002) – this is an empirically testable hypothesis. The most direct test would be the reevaluation of existing drug trial data from the geographic perspective of 'neighborhood effects', since, within the US, as a result of American Apartheid, 'race is place'.

While most drug trials are stratified by the usual population divisions of age, sex, race/ethnicity, income/education, and so on, they are seldom if ever examined from a formally geographic perspective. Detection of neighborhood effects in the spectrum of drug efficacy, compliance, and adverse reactions, would be a considerable advance. Changes in the pattern of such neighborhood effects over time, a likely consequence of diffusing social disintegration in the US, would serve to monitor the sociogeographic spread of drug failure and decay of the therapeutic alliance. This is likely to follow the classic mode of hierarchical hopscotch from major metropolitan regions like New York and Los Angeles, to smaller ones, then from central city to suburbs, and finally along social networks (e.g., Wallace and Wallace, 1997).

A second, and related line of work, would see development of a short preclinical interview instrument exploring (1) patterns of cross-sectional psychosocial stress and (2) geographic history of residence, and testing whether such an instrument would account for significant variance in drug efficacy, compliance, and adverse reactions. While this would certainly not substitute for programs of reform to ameliorate diffusing patterns of pathogenic social hierarchy and social disintegration, it might nonetheless prove of some individual-level value in adjusting drug regimens.

A third extension would involve expansion of this analysis to medical intervention beyond simple drug treatments, for example surgical outcomes and the results of psychosocial interventions. These too should show neighborhood effects indicating spatially and socially diffusing structured psychosocial stress according to the classic pattern.

A simple interview protocol would likely account for at least as much variance in prediction of treatment efficacy, compliance, and adverse reactions as more expensive and difficult genetic testing. It is even (remotely) possible that the two techniques could be used in parallel to provide a reasonably full characterization of the version of Lewontin's 'triple helix' which defines therapeutic response, a significant step toward a truly biological medicine far more in concert with the realities of human hypersociality than current simplistic and ideologically driven genetic determinism.

On a more somber note, however, while some individual-level tuning of conventional medical treatment does seem possible, it is painfully obvious that no medical system can ameliorate significant population-level health disparities in the absence of aggressive affirmative action to redress both the persisting burdens of history and current policies driving pathogenic hierarchy and social disintegration. This being said, it seems obvious that the economic well-being of the drug industry, if not precisely the health of the populations it services, will depend on a fuller understanding of the sociogeographic diffusion of drug failure, compliance failure, and adverse reactions. Pharmacogenetic theories of race constitute a significant barrier to such understanding.

1.9 References

Adams F. 2003, The informational turn in philosophy, *Minds and Machines*, 13:471-501.

Anfinsen, C., 1973, Principles that govern the folding of protein chains, *Science*, 181:223-230.

Ash, 1990, *Information Theory*, Dover Publications, New York.

Atlan H. and I. Cohen, 1998, Immune information, self-organization and meaning, *International Immunology*, 10:711-717.

Barkow J., L. Cosmides and J. Tooby, eds., 1992, *The Adapted Mind: Biological Approaches to Mind and Culture*, University of Toronto Press.

Beck C. and F. Schlogl, 1995, *Thermodynamics of Chaotic Systems*, Cambridge University Press.

Binney J., N. Dowrick, A. Fisher, and M. Newman, 1986, *The theory of critical phenomena*, Clarendon Press, Oxford, UK.

Bjorntorp P., 2001, Do stress reactions cause abdominal obesity and comorbidities? *Obesity Reviews*, 2:73-86.

Bonner J., 1980, *The evolution of culture in animals*, Princeton University Press, Princeton, NJ.

Braun L., 2002, Race, ethnicity, and health: can genetics explain disparities?, *Perspectives in Biological Medicine*, 45:159-74.

Brock C., and S. Salinsky, 1993, Empathy: an essential skill for understanding the physician-patient relationship in clinical practice, *Family Medicine*, 25:245-248.

Burroughs V., R. Maxey, and R. Levy, 2002, Racial and ethnic differences in response to medicines: towards individualized pharmaceutical treatment, *Journal of the National Medical Association*, 94(10 Suppl.):1-26. Cohen I., 1992, The cognitive principle challenges clonal selection, *Immunology Today*, 13:441-444.

Cohen I., 2000, *Tending Adam's Garden: Evolving the Cognitive Immune Self*, Academic Press, New York.

Cover T., and J. Thomas, 1991, *Elements of Information Theory*, John Wiley Sons, New York.

Dehaene S. and L Naccache, 2001, Towards a cognitive neuroscience of consciousness: basic evidence and a workspace framework, *Cognition* 79:1-37.

Damasio A., 1998, Emotion in the perspective of an integrated nervous system, *Brain Research Reviews*, 26:83-86.

Dimitrov A., and J. Miller, 2001, Neural coding and decoding: communication channels and quantization, *Computation and Neural Systems*, 12:441-472.

Downing R., and K. Rickels, 1982, The impact of favorable and unfavorable life events on psychotropic drug response, *Psychopharmacology*, 78:97-100.

Dretske F., 1981, *Knowledge and the flow of information*, MIT Press, Cambridge, MA.

Dretske F., 1988, *Explaining behavior*, MIT Press, Cambridge, MA.

Dretske F., 1992, What isn't wrong with fold psychology, *Metaphilosophy*, 23:1-13.

Dretske F., 1993, Mental events as structuring causes of behavior. In *Mental causation*, (ed. A. Mele and J. Heil), pp. 121-136. Oxford University Press.

Dretske F., 1994, The explanatory role of information, *Philosophical Transactions of the Royal Society, A*, 349:59-70.

Durham W., 1991, *Coevolution: Genes, Culture, and Human Diversity*, Stanford University Press, Palo Alto, CA.

Eldredge N., 1985, *Time Frames: The Rethinking of Darwinian Evolution and the Theory of Punctuated Equilibria*, Simon and Schuster, New York.

Exner D, D. Dries, M. Domanski, and J. Cohn, 2001, Lesser response to angiotensen converting enzyme therapy in black as compared with white patients with left ventrical dysfunction, *New England Journal of Medicine*, 344:1351-1357.

Fanon F., 1966, *The Wretched of the Earth*, Evergreen Press, New York.

Feynman R., 1996, *Feynman Lectures on Computation*, Addison-Wesley, Reading, MA.

Forlenza M. and A. Baum, 2000, Psychosocial influences on cancer progression: alternative cellular and molecular mechanisms, *Current Opinion in Psychiatry*, 13:639-645.

Gilbert P., 2001, Evolutionary approaches to psychopathology: the role of natural defenses, *Australian and New Zealand Journal of Psychiatry*, 35:17-27.

Glazebrook, J.F., and R. Wallace, 2009a, Small worlds and red queens in the global workspace, *Cognitive Systems Research*, 10:333-365.

Glazebrook, J.F., and R. Wallace, 2009b, Rate distortion manifolds as models for cognitive information, *Informatica*, 33:309-345.

Gould S., 2002, *The Structure of Evolutionary Theory*, Harvard University Press, Cambridge, MA.

Gordon A., 200, Cultural identity and illness: Fulani views, *Culture, Medicine and Psychiatry* 24:297-330.

Grossman A., J. Churchill, B. McKinney, I. Kodish, S. Otte, and W. Greenough, 2003, Experience effects on brain development: possible contributions to psychopathology, *Journal of Child Psychology and Psychiatry*, 44:33-63.

Grossman Z., 1989, The concept of idiotypic network: deficient or premature? In: H. Atlan and IR Cohen, (eds.), *Theories of Immune Networks*, Springer Verlag, Berlin, p. 3852.

Grossman Z., 1992a, Contextual discrimination of antigens by the immune system: towards a unifying hypothesis, in: A. Perelson and G. Weisbch, (eds.) *Theoretical and Experimental Insights into Immunology*, Springer Verlag, p. 7189.

Grossman Z., 1992b, *International Journal of Neuroscience*, 64:275.

Grossman Z., 1993, Cellular tolerance as a dynamic state of the adaptable lymphocyte, *Immunology Reviews*, 133:45-73.

Grossman Z., 2000, Round 3, *Seminars in Immunology*, 12:313-318.

Haller J., and J. Halasz, 2000, Effects of two acute stressors on the anxiolytic efficacy of chlordiazepoxide, *Psychopharmacology*, 151: 1-6.

Haller J., 2001, The link between stress and the efficacy of anxiolytics. A new avenue of research, *Physiology and Behavior*, 73: 337-342.

Langer N., 1999, Culturally competent professionals in therapeutic alliances enhance patient compliance, *Journal of Health Care for the Poor and Underserved*, 10:19-26.

Lazarou J., B. Pomeranz, and P. Correy, 1998, Incidence of adverse drug reactions in hospitalized patients–a meta-analysis of prospective studies, *Journal of the American Medical Association*, 279:1200-1205.

Lewontin R., 2000, *The Triple Helix: gene, organism, and environment*, Harvard University Press.

Link B. and J. Phelan, 2000, Evaluating the fundamental cause explanation for social disparities in health, in *Handbook of Medical Sociology*, Fifth Edition, C. Bird, P. Conrad, and A. Fremont (eds.), Prentice-Hall, NJ.

Massey D., 1990, American apartheid: segregation and the making of the underclass, *American Journal of Sociology*, 96:329-357.

Massey D., and N. Denton, 1993, *American Apartheid*, Harvard University Press.

Memmi A., 1967, *The Colonizer and the Colonized*, Beacon Press, Boston.

Memmi A., 1969, *Dominated Man*, Beacon Press, Boston.

Miura T., R. Kojima, M. Mizutani, Y. Shiga, F. Takatsu, and Y. Suzuki, 2001, Effect of digoxin noncompliance on hospitalization and mortality in patients with heart failure in long-term therapy: a prospective cohort study, *European Journal of Clinical Pharmacology*, 57:77-83.

Modiano D., V. Petrarca, B. Sirma, I. Nebie, D. Diallo, F. Esposito and M. Coluzzi, 1996, Different response to *Plasmodium falciparum* malaria in West African sympatric ethnic groups, *Proceedings of the National Academy of Sciences* 93:13206-13211.

Modiano D., G. Luoni, V. Petrarca, B. Sodiomon Sirima, M. De Luca, J. Simpore, M. Coluzzi, J. Bodmer and G. Modiano, 2001a, HLA class I in three West African ethnic groups: genetic distances from sub-Saharan and Caucasoid populations, *Tissue Antigen* 57:128-137.

Modiano D., G. Luoni, B. Sirima, A. Lanfrancotti, V. Petrarca, F. Cruciani, J. Simpore, B. Ciminelli, E. Foglietta, P. Grisanti, I. Bianco, G. Modiano and M. Coluzzi, 2001b, The lower susceptibility to Plasmodium falciparum malaria of Fulani of Burkina Faso (west Africa) is associated with low frequencies of classic malaria-resistance genes, *Transactions of the Royal Society of Tropical Hygiene and Medicine* 95:149-152.

Modiano D., V. Petrarca, B. Sirima, I. Nebie, G. Luoni, F. Esposito and M. Coluzzi, 1998, Baseline immunity of the population and impact of insecticide-treated curtains on malaria infection, *American Journal of Tropical and Medical Hygiene* 59:336-340.

Nunney L., 1999, Lineage selection and the evolution of multistage carcinogenesis, *Proceedings of the Royal Society, B*, 266:493-498.

Osmond C. and D. Barker, 2000, *Environmental Health Perspectives*, 108, Suppl. 3:545-553.

Pappas G., 1989, *The Magic City*, Cornell University Press, Ithaca, NY.

Park H., S. Amari, and K. Fukumizu, 2000, Adaptive natural gradient learning algorithms for various stochastic models, *Neural Networks*, 13:755-765.

Pielou E.C., 1977, *Mathematical Ecology*, John Wiley and Sons, New York.

Pirmohamed M., D. Naisbitt, F. Gordon, and B.K. Park, 2002, The danger hypothesis–potential role in idiosyncratic drug reaction, *Toxicology*, 181-182:55-63.

Podolsky S. and A. Tauber, 1997, *The generation of diversity: Clonal selection theory and the rise of molecular biology*, Harvard University Press.

Rau H. and T. Elbert, 2001, Psychophysiology of arterial baroreceptors and the etiology of hypertension, *Biological Psychology*, 57:179-201.

Richerson P. and R. Boyd, 1995, The evolution of human hypersociality, Paper for Rindberg Castle Symposium on Ideology, Warfare, and Indoctrination, (January, 1995), and HBES meeting, 1995.

Rojdestvensky I. and M. Cottam, 2000, Mapping of statistical physics to information theory with applications to biological systems, *Journal of Theoretical Biology*, 202:43-54.

Singh-Manoux A., N. Adler, and M. Marmot, 2003, Subjective social status: its determinants and its association with measures of ill-health in the Whitehall II study, *Social Science and Medicine*, 56:1321-1333.

Tauber A., 1998, Conceptual shifts in immunology: Comments on the 'two-way paradigm'. In K. Schaffner and T. Starzl (eds.), Paradigm Changes in Organ Transplantation, *Theoretical Medicine and Bioethics*, 19:457-473.

Thayer J., and R. Lane, 2000, A model of neurovisceral integration in emotion regulation and dysregulation, *Journal of Affective Disorders*, 61:201-216.

Thayer J., and B. Friedman, 2002, Stop that! Inhibition, sensitization, and their neurovisceral concomitants, *Scandinavian Journal of Psychology*, 43:123-130.

Timberlake W., 1994, Behavior systems, associationism, and Pavolvian conditioning. *Psychonomic Bulletin*, Rev. 1:405-420.

Wallace D., 1994, The resurgence of tuberculosis in New York City: a mixed hierarchically and spatially diffused epidemic, *American Journal of Public Health*, 84:1000-1002.

Wallace D., and R. Wallace, 1998, *A Plague on Your Houses*, Verso Publications, New York.

Wallace D. and R. Wallace, 2000, Life and death in Upper Manhattan and the Bronx: toward an evolutionary perspective on catastrophic social change, *Environment and Planning A*, 32:1245-1266.

Wallace R., M. Fullilove and A. Flisher, 1996, AIDS, violence and behavioral coding: information theory, risk behavior, and dynamic process on core-group sociogeographic networks, *Social Science and Medicine*, 43:339-352.

Wallace R., and Wallace D., 1997, Community marginalization and the diffusion of disease and disorder in the United States, *British Medical Journal*, 314:1341-1345.

Wallace R. and D. Wallace, 1997, The destruction of US minority urban communities and the resurgence of tuberculosis: ecosystem dynamics of the white plague in the de-developing world, *Environment and Planning A*, 29:269-291.

Wallace R., D. Wallace, and H. Andrews, 1997, AIDS, tuberculosis, violent crime and low birthweight in eight US metropolitan regions, *Environment and Planning A*, 29:525-555.

Wallace R. and R.G. Wallace, 1998, Information theory, scaling laws and the thermodynamics of evolution, *Journal of Theoretical Biology*, 192:545-559.

Wallace R., and R.G. Wallace, 1999, Organisms, organizations, and interactions: an information theory approach to biocultural evolution, *BioSystems*, 51:101-119.

Wallace R. and R. Fullilove, 1999, Why simple regression models work so well describing risk behaviors in the USA, *Environment and Planning A*, 31:719-734.

Wallace R., D. Wallace, J. E. Ullmann, and H. Andrews, 1999, Deindustrialization, inner-city decay, and the hierarchical diffusion of AIDS in the USA: how neoliberal and cold war policies magnified the ecological niche for emerging infections and created a national security crisis, *Environment and Planning A*, 31:113-139.

Wallace, R., and M. Fullilove, 2007, *Collective Consciousness and its Discontents: Institutional distributed cognition, racial policy, and public health in the United States*, Springer, New York.

Wallace R. and R.G. Wallace, 2002, Immune cognition and vaccine strategy: beyond genomics, *Microbes and Infection*, 4:521-527.

Wallace, R., and D. Wallace, 2008, Punctuated equilibrium in statistical models of generalized coevolutionary resilience: how sudden ecosystem transitions can entrain both phenotype expression and Darwinian selection, *Transactions on Computational Biology IX*, LNBI 5121:23-85.

Wallace, R., and D. Wallace, 2009, Code, context, and epigenetic catalysis in gene expression, *Transactions on Computational Biology XI*, LNABI 5750:283-334.

Wallace, R., R.G. Wallace, and D. Wallace, 2009, *Farming Human*

Pathogens: Ecological resilience and evolutionary process, Springer, New York.

Wallace, R., and D. Wallace, 2010, *Gene Expression and its Discontents: The social production of chronic disease*, Springer, New York.

Wallace R., 2000, Language and coherent neural amplification in hierarchical systems: renormalization and the dual information source of a generalized stochastic resonance, *International Journal of Bifurcation and Chaos*, 10:493-502.

Wallace R., 2002a, Immune cognition and vaccine strategy: pathogenic challenge and ecological resilience, *Open Systems and Information Dynamics*, 9:51-83.

Wallace R., 2002b, Adaptation, punctuation and rate distortion: non-cognitive 'learning plateaus' in evolutionary process, *Acta Biotheoretica*, 50:101-116.

Wallace R., 2003, Systemic lupus erythematosus in African-American women: cognitive physiological modules, autoimmune disease, and pathogenic social hierarchy, *Advances in Complex Systems*, 6:599-629.

Wallace, R., 2005, *Consciousness: A mathematical treatment of the neuronal global workspace model*, Springer, New York.

Wallace R., D. Wallace, and R.G. Wallace, 2003a, Toward cultural oncology: the evolutionary information dynamics of cancer, *Open Systems and Information Dynamics*, 10:159-181.

Wallace R., R.G. Wallace, and D. Wallace, 2003c, Coronary heart disease, chronic inflammation, and pathogenic social hierarchy: a biological limit to possible reductions in morbidity and mortality, *Journal of the National Medical Association*, 96:609-619.

Wallace R. and D. Wallace, 2005, Structured psychosocial stress and the US obesity epidemic, *Journal of Biological Systems*, 13:363-384.

Wallace R., D. Wallace, and M. Fullilove, 2003e, Community lynching and the US asthma epidemic.
http://cogprints.soton.ac.uk/archive/00002757/

Wallace R. and R.G. Wallace, 2004, Do reductionist cures select for holistic diseases? Adaptive chronic infection, structured stress, and medical magic bullets, *BioSystems*, 77:93-108.

Wallace R., 2004, Comorbidity in psychiatric and chronic physical diseae: autocognitive developmental disorders of structured psychosocial stress, *Acta Biotheoretica*, 52:71-93.

Wilkinson R., 1996, *Unhealthy Societies: the afflictions of inequality*, Routledge, London and New York.

Wilson, K., 1971, Renormalization group and critical phenomena. I Renormalization group and the Kadanoff scaling picture, *Physical Review B*, 4:3174-3183.

Wright R., M. Rodriguez, and S. Cohen, 1998, Review of psychosocial stress and asthma: an integrated biopsychosocial approach, *Tho-*

rax, 53:1066-1074.

Chapter 2

Heritability of complex disease

2.1 Introduction

2.1.1 Mental disorders and culture

Human mental disorders are not well understood, and their study provides a difficult example for theories of disease heritability. Official classifications as the *Diagnostic and Statistical Manual of Mental Disorders - fourth edition*, (DSM-IV, 1994), the standard descriptive nosology in the US, have even been characterized as 'prescientific' by P. Gilbert (2001) and others. Johnson-Laird et al. (2006) claim that current knowledge about psychological illnesses is comparable to the medical understanding of epidemics in the early 19th century. Physicians realized then that cholera, for example, was a specific disease, which killed about a third of the people whom it infected. What they disagreed about was the cause, the pathology, and the communication of the disease. Similarly, according to Johnson-Laird et al., most medical professionals these days realize that psychological illnesses occur (cf. DSMIV), but they disagree about their cause and pathology. Notwithstanding DSMIV, Johnson-Laird et al. doubt whether any satisfactory a priori definition of psychological illness can exist because it is a matter for theory to elucidate.

Atmanspacher (2006) argues that formal theory of high level cognitive process is itself at a point similar to that of physics 400 years ago, in that the basic entities, and the relations between them, have yet to be delineated.

More generally, simple arguments from genetic determinism regarding mental disorders fail, in part because of a draconian population bottleneck that, early in our species' history, resulted in an overall genetic diversity less than that observed within and between contemporary chimpanzee subgroups. Manolio et al. (2009) describe this conundrum more generally in terms of 'finding the missing heritability of complex diseases'. They observe, for example, that at least 40 loci have been associated with human height, a classic complex trait with an estimated heritability of about 80 %, yet they explain only about 5 % of phenotype variance despite studies of tens of thousands of people. This result, they find, is typical across a broad range of supposedly heritable diseases, and call for extending beyond current genome-wide assoication approaches to illuminate the genetics of complex diseases and enhance its potential to enable effective disease prevention or treatment.

Arguments from psychosocial stress fare better (e.g., Brown et al., 1973; Dohrenwend and Dohrenwend, 1974; Eaton, 1978), particularly for depression (e.g., Risch et al., 2009), but are affected by the apparently complex and contingent developmental paths determining the onset of schizophrenia, dementias, psychoses, and so forth, some of which may be triggered in utero by exposure to infection, low birthweight, or other functional teratogens.

P. Gilbert suggests an extended evolutionary perspective, in which evolved mechanisms like the 'flight-or-fight' response are inappropriately excited or suppressed, resulting in such conditions as anxiety or post traumatic stress disorders. Nesse (2000) suggests that depression may represent the dysfunction of an evolutionary adaptation which down-regulates foraging activity in the face of unattainable goals.

Kleinman and Good, however, (1985, p. 492) outline something of the cross cultural subtleties affecting the study of depression that seem to argue against any simple evolutionary or genetic interpretation. They state that, when culture is treated as a constant, as is common when studies are conducted in our own society, it is relatively easy to view depression as a biological disorder, triggered by social stressors in the presence of ineffective support, and reflected in a set of symptoms or complaints that map back onto the biological substrate of the disorder. However, they continue, when culture is treated as a significant variable, for example, when the researcher seriously confronts the world of meaning and experience of members of non-Western societies, many of our assumptions about the nature of emotions and illness are cast in sharp relief. Dramatic differences are found across cultures in the social organization, personal experience, and consequences of such emotions as sadness, grief, and anger, of behaviors such as withdrawal or aggression, and of psychological characteristics such as passivity and helplessness or the resort to altered states of consciousness. They are organized differently as psychological realities, communicated in a wide range of idioms, related to quite varied local contexts of power relations, and are interpreted, evaluated, and responded to as fundamentally different meaningful realities. Depressive illness and dysphoria are thus not only interpreted differently in non-Western societies and across cultures; they are *constituted* as fundamentally different forms of social reality.

Since publication of that landmark study, a number of comprehensive overviews support its conclusions, for example Bebbington (1993), Jenkins, Kleniman and Good (1990), *Journal of Clinical Psychiatry* (Supplement 13), and Manson (1995). As Marsella (2003) writes, it is now clear that cultural variations exist in the areas of meaning, perceived causes, onset patterns, epidemiology, symptom expression, course, and outcome, variations having important implications for understanding clinical activities including conceptualization, assessment, and therapy.

Kleinman and Cohen (1997) argue forcefully that several myths have become central to psychiatry. The first is that the forms of mental illness everywhere display similar degrees of prevalence. The second is an excessive adherence to a principle known as the

pathogenic/pathoplastic dichotomy, which holds that biology is responsible for the underlying structure of a malaise, whereas cultural beliefs shape the specific ways in which a person experiences it. The third myth maintains that various unusual culture-specific disorders whose biological bases are uncertain occur only in exotic places outside the West. In an effort to base psychiatry in 'hard' science and thus raise its status to that of other medical disciplines, psychiatrists have narrowly focused on the biological underpinnings of mental disorders while discounting the importance of such 'soft' variables as culture and socioeconomic status.

Heine (2001) describes an explicit cultural psychology that views the person as containing a set of biological potentials interacting with particular situational contexts that constrain and afford the expression of various constellations of traits and patterns of behavior. He says that, unlike much of personality psychology, cultural psychology focuses on the constraints and affordances inherent to the cultural environment that give shape to those biological potentials. Human nature, from this perspective, is seen as emerging from participation in cultural worlds, and of adapting oneself to the imperatives of cultural directives, meaning that our nature is ultimately that of a cultural being.

Heine describes how cultural psychology does not view culture as a superficial wrapping of the self, or as a framework within which selves interact, but as something that is intrinsic to the self, so that without culture there is no self, only a biological entity deprived of its potential. Individual selves, from Heine's perspective, are inextricably grounded in a configuration of consensual understandings and behavioral customs particular to a given cultural and historical context, so that understanding the self requires an understanding of the culture that sustains it. Heine argues, then, that the process of becoming a self is contingent on individuals interacting with, and seizing meanings from, the cultural environment.

Heine warns that the extreme nature of American individualism means that a psychology based on late 20th century American research not only stands the risk of developing models that are particular to that culture, but of developing an understanding of the self that is peculiar in the context of the world's cultures.

Indeed, as Norenzayan and Heine (2005) point out, for the better part of a hundred years, a considerable controversy has raged within anthropology regarding the degree to which psychological and other human universals do, in fact, actually exist independent of the particularities of culture.

Arnett (2008), in a paper provocatively titled *The Neglected 95 %*, similarly argues that US psychological research focuses too narrowly on Americans, who comprise less than 5 percent of the world's population, and on perhaps another 7 percent in Western countries. He asserts that the majority of the world's population lives in under vastly different conditions, underlying doubts of how representative American psychological research can be, and finds the narrowness of American research to be a consequence of a focus on a philosophy of science that emphasizes fundamental processes and ignores or strips away cultural context.

Henrich et al. (2009), in a wide-ranging review, extend the considerations of Norenzayan and Heine, finding that Western, educated, industrialized and democratic (WEIRD) subjects, across domains of visual perception, fairness, categorization, spatial cognition, memory, moral reasoning, and self-concepts, constitute frequent outliers compared with the rest of the species. They conclude that addressing questions of *human* psychology will require tapping broader subject pools.

As Durham (1991) and Richerson and Boyd (2004) explore at some length, humans are endowed with two distinct but interacting heritage systems: genes and culture. Durham (1991), for example, writes that genes and culture constitute two distinct but interacting systems of information inheritance within human populations and information of both kinds has influence, actual or potential, over behaviors, which creates a real and unambiguous symmetry between genes and phenotypes on the one hand, and culture and phenotypes on the other. Genes and culture, in his view, are best represented as two parallel lines or tracks of hereditary influence on phenotypes.

Both genes and culture can be envisioned as generalized languages in that they have recognizable 'grammar' and 'syntax', in the sense of Wallace (2005) and Wallace and Wallace (2008, 2009).

More recent work has identified epigenetic heritage mechanisms involving such processes as environmentally-induced gene methylation, that can have strong influence across several generations (e.g., Jablonka and Lamb, 1995, 1998; Jabolonka, 2004), and are the subject of intense current research.

There are, it seems, two powerful heritage mechanisms in addition to the genetic where one may perhaps find the 'missing heritability of complex diseases' that Manolio et al. seek.

Here, however, we are particularly interested in the phenotypes of madness, and in their relations to genes, culture, and environment.

2.1.2 Two case histories

Gene-environment interaction

Much recent work in American biological psychiatry has emphasized the search for gene-environment interactions. Caspi and Moffitt (2006), for example, claim that such interactions occur when the effect of exposure to an environmental pathogen on a person's health is conditional on his or her genotype. The first evidence that genotype moderates the capacity of an environmental risk to bring about mental disorders was, according to them, reported in 2002, (Caspi et al., 2002), in a study of the role of genotype in the cycle of violence in maltreated children. Caspi and Moffitt (2006) claim that the gene-environment interaction approach brings opportunities for extending the range and power of neuroscience by introducing opportunities for collaboration between experimental neuroscience and research on gene-environment interactions. Successful collaboration can, in their view, solve the biggest mystery of human psychopathology: how does an environmental factor, external to the person, get inside the nervous system and alter its elements to generate the symptoms of a disordered mind? Concentrating the considerable resources of neuroscience and gene-environment interaction on this question will, they claim, bring discoveries that advance the understanding of mental disorders, and increase the potential to control and prevent them.

One of the most cited of recent studies of gene-environment interactions is, indeed, the work of Caspi et al. (2003), who found that genetic variation in the promoter region of the serotonin transporter gene

(5-HTTLPR;[OMIM182138]), in interaction with stressful life events, contributes to a predisposition to major depression. As Risch et al. (2009) put it, this result was striking and potentially paradigm shifting because numerous previous studies of this same polymorphism, without examining environmental risk factors or life events, had not consistently shown either a strong or replicated association with depression. A subsequent meta-analysis was conducted by Risch et al. (2009) that combined data from some 14 studies having a total of 14,250 participants, some 1769 of whom met criteria for depression.

Risch et al. state that most of the participants were white, except for a multiethnic sample in one study, and an Asian sample in another. Contrary to the results of Caspi et al. (2003), they found no evidence that the serotonin transporter genotype alone, or in interaction with stressful life events, is associated with an elevated risk of depression.

The Asian study, by J. Kim et al. (2006), involved 732 Korean community residents ages 65+, a fair number indeed. Some 88 percent at baseline did not meet criteria for depression. Kim et al., in contrast with Risch et al., in spite of using 'standard' instruments for both measures of depression and life events (translated into Korean), found a strong statistical trend suggesting that environmental risk of depression is indeed modified by at least two genes, and that gene-environment interactions are found even into old age.

Given the scathing analyses by Arnett, Heine, and Henrich et al., the bitter conflict between the results of Caspi et al. (2003) and Risch et al. (2009) is in serious danger of becoming simply a culture-bound tempest in a distinctly American teapot.

Gene-culture interaction

The necessity for the inclusion of culture in the operation of fundamental psychological phenomena is emphasized by the observations of Nisbett et al. (2001), and others, following the tradition of Markus and Kitayama (1991), regarding profound differences in basic perception between test subjects of Southeast Asian and Western cultural heritage across an broad realm of experiments. East Asian perspectives are characterized as *holistic* and Western as *analytic*. Nisbett et al. (2001) find:

(1) Social organization directs attention to some aspects of the perceptual field at the expense of others.

(2) What is attended to influences metaphysics.

(3) Metaphysics guides tacit epistemology, that is, beliefs about the nature of the world and causality.

(4) Epistemology dictates the development and application of some cognitive processes at the expense of others.

(5) Social organization can directly affect the plausibility of metaphysical assumptions, such as whether causality should be regarded as residing in the field vs. in the object.

(6) Social organization and social practice can directly influence the development and use of cognitive processes such as dialectical vs. logical ones.

Nisbett et al. (2001) conclude that tools of thought embody a culture's intellectual history, that tools have theories built into them, and that users accept these theories, albeit unknowingly, when they use these tools.

More recently, Masuda and Nisbett (2006) examined cultural variations in change blindness, a phenomenon related to inattentional blindness, and found striking differences between Western and East Asian subjects. They presented participants with still photos and with animated vignettes having changes in focal object information and contextual information. Compared to Americans, East Asians were more sensitive to contextual changes than to focal object changes. These results, they conclude, suggest that there can be cultural variation in what may seem to be basic perceptual processes.

H. Kim et al. (2010) have extended this line of work to examine the interaction between genes and culture as determinants of individuals' locus of attention. As the serotonin (5-HT) system has been associated with attentional focus and the ability to adapt to changes in reinforcement, they examined the serotonin 1A receptor polymorphism (5-HTR1A). Koreans and European Americans were geneotyped and reported their chronic locus of attention. They found a significant interaction between 5-HTR1A and culture in the locus of attention. Koreans reported attending to the field more than European Americans, and this cultural difference was moderated by 5-HTR1A. There was a linear pattern such that those homozygous for the G allele, which is associated with reduced ability to adapt to changes in reinforcement, more strongly endorsed the culturally reinforced mode of thinking than those homozygous for the C allele, with those heterozygous in the middle. Kim et al. claim that their findings suggest that the same genetic predisposition can result in divergent psychological outcomes, depending on an individual's cultural context.

The sample used in this study included 149 Korean and 140 European subjects. Given the problems with the Caspi et al. work, it is clear that replication across larger samples will be needed.

That being said, the results of H. Kim et al. do indeed underline the necessity of expanding work on psychiatric disorders to gene-culture-environment interactions. It seems likely, however, that, overall, culture-environment interaction effects will predominate. Nonetheless, the effects of genetic structure on that interaction might well provide important insights as to etiology and possible treatment.

2.1.3 Global broadcast models

Recent research on schizophrenia, dyslexia, and autism, broadly supports a 'brain connectivity' model for these disorders that is of considerable interest from the perspective of global workspace/global broadcast models of consciousness that are the foundation of our work (e.g., Baars, 1989; Wallace, 2005), since large-scale brain connectivity is essential for the operation of consciousness, a principal, and very old, evolutionary adaptation (e.g., Wallace and Wallace, 2009).

For example, Burns et al. (2003), on the basis of sophisticated diffusion tensor magnetic resonance imaging studies, argue that schizophrenia is a disorder of large-scale neurocognitive networks rather than specific regions, and that pathological changes in the disorder should be sought at the supra-regional level. Both structural and functional abnormalities in frontoparietal networks have been described and may constitute a basis for the wide range of cognitive functions impaired in the disorder, such as selective attention, language processing and attribution of agency.

Silani et al. (2005) find that, for dyslexia, altered activation observed within the reading system is associated with altered density of grey and white matter of specific brain regions, such as the left middle and inferior temporal gyri and left arcuate fasciculus. This supports the view that dyslexia is associated with both local grey matter dysfunction and with altered larger scale connectivity among phonological/reading areas.

Villalobos et al. (2005) explore the hypothesis that large-scale abnormalities of the dorsal stream and possibly the mirror neuron system, may be responsible for impairments of joint attention, imitation, and secondarily for language delays in autism. Their empirical study showed that those with autism had significantly reduced connectivity with bilateral inferior frontal area 44, which is compatible with the hypothesis of mirror neuron defects in autism. More generally, their results suggest that dorsal stream connectivity in autism may not be fully functional.

Courchesne and Pierce (2005) suggest that, for autism, connectivity within the frontal lobe is excessive, disorganized, and inadequately selective, whereas connectivity between frontal cortex and other systems is poorly synchronized, weakly responsive and information impoverished. Increased local but reduced long-distance cortical-cortical reciprocal activity and coupling would impair the fundamental frontal function of integrating information from widespread and diverse systems and providing complex context-rich feedback, guidance and control to lower-level systems.

Coplan (2005) has observed a striking pattern of excessive frontal lobe self-connectivity in certain cases of anxiety disorder, and Coplan et al. (2005) find that maternal stress can affect long-term hippocampal neurodevelopment in a primate model.

As stated, brain connectivity is the *sine qua non* of Global Workspace and Broadcast models of individual mental function including consciousness (e.g., Baars, 1989; Wallace, 2005), and further analysis suggests that these disorders cannot be fully understood in the absence of a functional theory of consciousness, and in particular, of a detailed understanding of the elaborate regulatory mechanisms which must have evolved over the past half billion years to ensure the stability of that most central and most powerful of adaptations. For humans, of course, one of the principal regulatory mechanisms for mental function is the embedding culture and culturally-mediated social interaction, in addition to culture's role as the second system of human heritage. As the evolutionary anthropologist Robert Boyd has put it, culture is as much a part of human biology as the enamel on our teeth (e.g., Richerson and Boyd, 2004).

Distortion of consciousness is not simply an epiphenomenon of the emotional dysregulation which many see as the 'real' cause of mental disorder. Like the pervasive effects of culture, distortion of consciousness lies at the heart of both the individual experience of mental disorder and the effect of it on the embedding of the individual within both social relationships and cultural or environmental milieu. Yet the experience of individual consciousness cannot be disentangled from social interaction and culture (Wallace, 2005). Distortion of a culturally-mediated consciousness in mental disorders inhibits both routine social interchange and the ability to meet internalized or expected cultural norms, a potentially destabilizing positive feedback. Distortion of consciousness profoundly affects the ability to learn new, or change old, skills in the face of changing patterns of threat or opportunity, perhaps the most critical purpose of the adaptation itself. Distortion of consciousness causing decoupling from social and cultural context is usually a threat to long-term individual survival, and those with mental disorders significantly affecting consciousness typically experience severely shortened lifespans.

Here we will expand recent explorations of a cognitive paradigm for gene expression (Wallace and Wallace, 2008, 2009) that incorporates the effects of surrounding epigenetic regulatory machinery as a kind of catalyst to include the effects of the embedding information source of human culture on the ontology of the human mind. The essential feature is that a cognitive process, including gene expression, can instantiate a dual information source that can interact with the generalized language of culture in which, for example, social interplay and the interpretation of socioeconomic and environmental stressors, involve complicated matters of symbolism and its grammar and syntax. These information sources interact by a crosstalk that, over the life course, determines the ontology of mind, including its manifold dysfunctions.

That is, contemporary American work on gene-environment interactions in psychiatry must be extended to the study of gene-culture-environment interactions. This is no small matter, and the probability models we present here are at the borders of current applied mathematics.

2.1.4   A cognitive paradigm for gene expression

A cognitive paradigm for gene expression has emerged in which contextual factors determine the behavior of what Cohen characterizes as a 'reactive system', not at all a deterministic, or even stochastic, mechanical process (e.g., Cohen, 2006; Cohen and Harel, 2007; Wallace and Wallace, 2008, 2009). The very different formal approaches are, however, all in the spirit of Maturana and Varela (1980, 1992) who understood the central role that cognitive process must play across a vast array of biological phenomena.

O'Nuallain (2008) has placed gene expression firmly in the realm of complex linguistic behavior, for which context imposes meaning, claiming that the analogy between gene expression and language production is useful both as a fruitful research paradigm and also, given the relative lack of success of natural language processing (nlp) by computer, as a cautionary tale for molecular biology. A relatively simple model of cognitive process as an information source permits use of Dretske's (1994) insight that any cognitive phenomenon must be constrained by the limit theorems of information theory, in the same sense that sums of stochastic variables are constrained by the Central Limit Theorem. This perspective permits a new formal approach to gene expression and its dysfunctions, in particular suggesting new and powerful statistical tools for data analysis that could contribute to exploring both ontology and its pathologies. Wallace and Wallace (2009, 2010) apply the perspective, respectively, to infectious and chronic disease. Here we extend the mathematical foundations of that work to examine the topological structures of development and developmental disorder, in the context of an embedding information source representing the compelling varieties of human culture.

This approach is consistent with the broad context of epigenetics and epigenetic epidemiology. In particular, Jablonka and Lamb (1995, 1998) argue that information can be transmitted from one generation to the next in ways other than through the base sequence of DNA. It can be transmitted through cultural and behavioral means in higher animals, and by epigenetic means in cell lineages. All of these transmission systems allow the inheritance of environmentally induced variation. Such Epigenetic Inheritance Systems are the memory systems that enable somatic cells of different phenotypes but identical genotypes to transmit their phenotypes to their descendants, even when the stimuli that originally induced these phenotypes are no longer present.

After a decade of research and debate, the epigenetic perspective has received much empirical confirmation (e.g., Backdahl et al. 2009; Turner, 2000; Jaenish and Bird, 2003; Jablonka, 2004).

Foley et al. (2009) argue that epimutation is estimated to be 100 times more frequent than genetic mutation and may occur randomly or in response to the environment. Periods of rapid cell division and epigenetic remodeling are likely to be most sensitive to stochastic or environmentally mediated epimutation. Disruption of epigenetic profile is a feature of most cancers and is speculated to play a role in the etiology of other complex diseases including asthma allergy, obesity, type 2 diabetes, coronary heart disease, autism spectrum disorders and bipolar disorders and schizophrenia.

Important work by Scherrer and Jost (2007a, b) that is similar to the approach of this paper explicitly invokes information theory in their extension of the definition of the gene to include the local epigenetic machinery, a construct they term the 'genon'. Their central point is that coding information is not simply contained in the coded sequence, but is, in their terms, *provided by* the genon that accompanies it on the expression pathway and controls in which peptide it will end up. In their view the information that counts is not about the identity of a nucleotide or an amino acid derived from it, but about the relative frequency of the transcription and generation of a particular type of coding sequence that then contributes to the determination of the types and numbers of functional products derived from the DNA coding region under consideration.

The proper formal tools for understanding phenomena that 'provide' information – that are information sources – are the Rate Distortion Theorem and its zero error limit, the Shannon-McMillan Theorem.

2.2 Models of development

The currently popular spinglass model of development (e.g., Ciliberti et al., 2007a, b) assumes that N transcriptional regulators, are represented by their expression patterns

$$\mathbf{S}(t) = [S_1(t), ..., S_N(t)]$$

(2.1)

at some time t during a developmental or cell-biological process and in one cell or domain of an embryo. The transcriptional regulators influence each other's expression through cross-regulatory and autoregulatory interactions described by a matrix $w = (w_{ij})$. For nonzero elements, if $w_{ij} > 0$ the interaction is activating, if $w_{ij} < 0$ it is repressing. w represents, in this model, the regulatory genotype of the system, while the expression state $\mathbf{S}(t)$ is the phenotype. These regulatory interactions change the expression of the network $\mathbf{S}(t)$ as time progresses according to a difference equation

$$S_i(t + \Delta t) = \sigma[\sum_{j=1}^{N} w_{ij} S_j(t)],$$

(2.2)

where Δt is a constant and σ a sigmodial function whose value lies in the interval $(-1, 1)$. In the spinglass limit σ is the sign function, taking only the values ± 1.

The regulatory networks of interest here are those whose expression state begins from a prespecified initial state $\mathbf{S}(0)$ at time $t = 0$ and converge to a prespecified stable equilibrium state $\mathbf{S}_\infty$. Such networks are termed *viable* and must necessarily be a very small fraction of the total possible number of networks, since most do not begin and end on the specified states. This 'simple' observation is not at all simple in our reformulation, although other results become far more accessible, as we can then invoke the asymptotic limit theorems of information theory.

The spinglass approach to development is formally similar to spinglass neural network models of learning by selection, e.g., as proposed by Toulouse et al. (1986) nearly a generation ago. Much subsequent work, summarized by Dehaene and Naccache (2001), suggests that such models are simply not sufficient to the task of understanding high level cognitive function, and these have been largely supplanted by complicated 'global workspace' concepts whose mathematical characterization is highly nontrivial (Atmanspacher, 2006).

Wallace and Wallace (2008, 2009) shift the perspective on development by invoking a cognitive paradigm for gene expression, following the example of the Atlan/Cohen model of immune cognition.

Atlan and Cohen (1998), in the context of a study of the immune system, argue that the essence of cognition is the comparison of a perceived signal with an internal, learned picture of the world, and then choice of a single response from a large repertoire of possible responses.

Such choice inherently involves information and information transmission since it always generates a reduction in uncertainty, as explained by Ash (1990, p. 21).

More formally, a pattern of incoming input – like the $\mathbf{S}(t)$ above – is mixed in a systematic algorithmic manner with a pattern of internal ongoing activity – like the (w_{ij}) above – to create a path of combined signals $x = (a_0, a_1, ..., a_n, ...)$ – analogous to the sequence of $\mathbf{S}(t + \Delta t)$ above, with, say, $n = t/\Delta t$. Each a_k thus represents some functional composition of internal and external signals.

This path is fed into a highly nonlinear decision oscillator, h, a 'sudden threshold machine', in a sense, that generates an output $h(x)$ that is an element of one of two disjoint sets B_0 and B_1 of possible system responses. Let us define the sets B_k as

$$B_0 \equiv \{b_0, ..., b_k\},$$

$$B_1 \equiv \{b_{k+1}, ..., b_m\}.$$

(2.3)

Assume a graded response, supposing that if $h(x) \in B_0$,the pattern is not recognized, and if $h(x) \in B_1$, the pattern has been recognized, and some action b_j, $k + 1 \leq j \leq m$ takes place.

The principal objects of formal interest are paths x triggering pattern recognition-and-response. That is, given a fixed initial state a_0, examine all possible subsequent paths x beginning with a_0 and leading to the event $h(x) \in B_1$. Thus $h(a_0, ..., a_j) \in B_0$ for all $0 < j < m$, but $h(a_0, ..., a_m) \in B_1$.

For each positive integer n, let $N(n)$ be the number of high probability grammatical and syntactical paths of length n which begin with some particular a_0 and lead to the condition $h(x) \in B_1$. Call such paths 'meaningful', assuming, not unreasonably, that $N(n)$ will be considerably less than the number of all possible paths of length n leading from a_0 to the condition $h(x) \in B_1$.

While the combining algorithm, the form of the nonlinear oscillator, and the details of grammar and syntax are all unspecified in this model, the critical assumption which permits inference of the necessary conditions constrained by the asymptotic limit theorems of information theory is that the finite limit

$$H \equiv \lim_{n \to \infty} \frac{\log[N(n)]}{n}$$

(2.4)

both exists and is independent of the path x.

Define such a pattern recognition-and-response cognitive process as *ergodic*. Not all cognitive processes are likely to be ergodic in this sense, implying that H, if it indeed exists at all, is path dependent, although extension to nearly ergodic processes seems possible (Wallace and Fullilove, 2008).

Invoking the spirit of the Shannon-McMillan Theorem, as choice involves an inherent reduction in uncertainty, it is then possible to define an adiabatically, piecewise stationary, ergodic (APSE) information source $\mathbf{X}$ associated with stochastic variates X_j having joint and conditional probabilities $P(a_0, ..., a_n)$ and $P(a_n|a_0, ..., a_{n-1})$ such that appropriate conditional and joint Shannon uncertainties satisfy the classic relations

$$H[\mathbf{X}] = \lim_{n \to \infty} \frac{\log[N(n)]}{n}$$

$$= \lim_{n \to \infty} H(X_n|X_0, ..., X_{n-1})$$

$$= \lim_{n \to \infty} \frac{H(X_0, ..., X_n)}{n+1}.$$

(2.5)

This information source is defined as *dual* to the underlying ergodic cognitive process.

Adiabatic means that the source has been parametrized according to some scheme, and that, over a certain range, along a particular piece, as the parameters vary, the source remains as close to stationary and ergodic as needed for information theory's central theorems to apply. *Stationary* means that the system's probabilities do not change in time, and *ergodic*, roughly, that the cross sectional means approximate long-time averages. Between pieces it is necessary to invoke various kinds of phase transition formalisms, as described more fully in Wallace (2005) or Wallace and Wallace (2008).

In the developmental vernacular of Ciliberti et al., we now examine paths in phenotype space that begin at some $\mathbf{S}_0$ and converge $n = t/\Delta t \to \infty$ to some other $\mathbf{S}_\infty$. Suppose the system is conceived at $\mathbf{S}_0$, and h represents (for example) reproduction when phenotype $\mathbf{S}_\infty$ is reached. Thus $h(x)$ can have two values, i.e., B_0 not able to reproduce, and B_1, mature enough to reproduce. Then $x = (\mathbf{S}_0, \mathbf{S}_{\Delta t}, ..., \mathbf{S}_{n\Delta t}, ...)$ until $h(x) = B_1$.

Structure is now subsumed *within the sequential grammar and syntax of the dual information source* rather than within the cross sectional internals of (w_{ij})-space, a simplifying shift in perspective.

This transformation carries computational burdens, as well as providing mathematical insight.

First, the fact that viable networks comprise a tiny fraction of all those possible emerges easily from the spinglass formulation simply because of the 'mechanical' limit that the number of paths from $\mathbf{S}_0$ to $\mathbf{S}_\infty$ will always be far smaller than the total number of possible paths, most of which simply do not end on the target configuration.

From the information source perspective, which inherently subsumes a far larger set of dynamical structures than possible in a spinglass model – not simply those of symbolic dynamics – the result is what Khinchin (1957) characterizes as the 'E-property' of a stationary, ergodic information source. This property allows, in the limit of infinitely long output, the classification of output strings into two sets:

[1] a very large collection of gibberish which does not conform to underlying (sequential) rules of grammar and syntax, in a large sense, and which has near-zero probability, and

[2] a relatively small 'meaningful' set, in conformity with underlying structural rules, having very high probability.

The essential content of the Shannon-McMillan Theorem is that, if $N(n)$ is the number of meaningful strings of length n, then the uncertainty of an information source X can be defined as

$H[X] = \lim_{n \to \infty} \log[N(n)]/n$,

that can be expressed in terms of joint and conditional probabilities. Proving these results for general stationary, ergodic information sources requires considerable mathematical machinery (e.g., Khinchin, 1957; Cover and Thomas, 1991; Dembo and Zeitouni, 1998).

Second, according to Ash (1990) information source uncertainty has an important heuristic interpretation in that we may regard a portion of text in a particular language as being produced by an information source. A large uncertainty means, by the Shannon-McMillan Theorem, a large number of 'meaningful' sequences. Thus given two languages with uncertainties H_1 and H_2 respectively, if $H_1 > H_2$, then in the absence of noise it is easier to communicate in the first language; more can be said in the same amount of time. On the other hand, it will be easier to reconstruct a scrambled portion of text in the second language, since fewer of the possible sequences of length n are meaningful.

Third, information source uncertainty is homologous with free energy density in a physical system, a matter having implications across a broad class of dynamical behaviors.

The free energy density of a physical system having volume V and partition function $Z(K)$ derived from the system's Hamiltonian – the energy function – at inverse temperature K is (e.g., Landau and Lifshitz 2007)

$$F[K] = \lim_{V \to \infty} -\frac{1}{K} \frac{\log[Z(K,V)]}{V}$$

$$= \lim_{V \to \infty} \frac{\log[\hat{Z}(K,V)]}{V},$$

(2.6)

where $\hat{Z} = Z^{-1/K}$.

The partition function for a physical system is the normalizing sum in an equation having the form

$$P[E_i] = \frac{\exp[-E_i/kT]}{\sum_j \exp[-E_j/kT]},$$

(2.7)

where E_i is the energy of state i, k a constant, and T the system temperature.

Feynman (2000), following the classic approach by Bennett (1988), who examined idealized machines using information to do work, concludes that *the information contained in a message is most simply measured by the free energy needed to erase it.*

Thus, according to this argument, source uncertainty is homologous to free energy density as defined above, i.e., from the similarity with the relation $H = \lim_{n \to \infty} \log[N(n)]/n$.

Ash's perspective then has an important corollary: If, for a biological system, $H_1 > H_2$, source 1 will require more metabolic free energy than source 2.

2.3 Tunable epigenetic catalysis

Following the direction of Wallace and Wallace (2009), incorporating the influence of embedding contexts – generalized epigenetic effects – is most elegantly done by invoking the Joint Asymptotic Equipartition Theorem (JAEPT) (Cover and Thomas, 1991). For example, given an embedding epigenetic information source, say Y, that affects development, then the dual cognitive source uncertainty $H[X]$ is replaced by a joint uncertainty $H(X,Y)$. The objects of interest then become the jointly typical dual sequences $z^n = (x^n, y^n)$, where x is associated with cognitive gene expression and y with the embedding epigenetic regulatory context. Restricting consideration of x and y to those sequences that are in fact jointly typical allows use of the information transmitted from Y to X as the splitting criterion.

One important inference is that, from the information theory 'chain rule' (Cover and Thomas, 1991), $H(X,Y) = H(X) + H(Y|X) \leq H(X) + H(Y)$, while there are approximately $\exp[nH(X)]$ typical X sequences, and $\exp[nH(Y)]$ typical Y sequences, and hence $\exp[n(H(x) + H(Y))]$ independent joint sequences, there are only

$$\exp[nH(X,Y)] \leq \exp[n(H(X) + H(Y))]$$

jointly typical sequences, so that the effect of the embedding context is to lower the relative free energy of a particular developmental channel. Equality occurs only for stochastically independent processes.

Thus the effect of epigenetic regulation is to channel development into pathways that might otherwise be inhibited by an energy barrier. Hence the epigenetic information source Y acts as a *tunable catalyst*, a kind of second order cognitive enzyme, to enable and direct developmental pathways. This result permits hierarchical models similar to those of higher order cognitive neural function that incorporate contexts in a natural way (e.g., Wallace and Wallace, 2008; Wallace and Fullilove, 2008). The cost of this ability to channel is the metabolic necessity of supporting two information sources, X and Y, rather than just Y itself.

This elaboration allows a spectrum of possible 'final' phenotypes, what S. Gilbert (2001) calls developmental or phenotype plasticity. Thus gene expression is seen as, in part, responding to environmental or other, internal, developmental signals.

Including the effects of embedding culture in the development of the human mind is, according to this formalism, quite straightforward: Consider culture as another embedding information source, Z, having source uncertainty $H(Z)$. Then the information chain rule becomes

$$H(X, Y, Z) \leq H(X) + H(Y) + H(Z)$$

(2.8)

and

$$\exp[nH(X, Y, Z)] \leq \exp[n(H(X) + H(Y) + H(Z))],$$

(2.9)

where, again, equality occurs only under stochastic independence.

In this model, following explicitly the direction indicated by Boyd, Kleinman and their colleagues, culture is seen here as an essential component of the catalytic epigenetic machinery that regulates the development of the human mind. This is not to say that the development of mind in other animals, particularly those that are highly social, does not undergo analogous regulation by larger scale structures of interaction. For human populations, however, social relations are themselves very highly regulated through an often strictly formalized cultural grammar and syntax.

2.4   Groupoid free energy

A formal equivalence class algebra can now be constructed by choosing different origin and end points $\mathbf{S}_0, \mathbf{S}_\infty$ and defining equivalence of two states by the existence of a high probability meaningful path connecting them with the same origin and end. Disjoint partition by equivalence class, analogous to orbit equivalence classes for dynamical systems, defines the vertices of the proposed network of cognitive dual languages, much enlarged beyond the spinglass example. We thus envision a network of metanetworks. Each vertex then represents a different equivalence class of information sources dual to a cognitive process. This is an abstract set of metanetwork 'languages' dual to the cognitive processes of gene expression and development.

This structure generates a groupoid, in the sense of Weinstein (1996). States a_j, a_k in a set A are related by the groupoid morphism if and only if there exists a high probability grammatical path connecting them to the same base and end points, and tuning across the various possible ways in which that can happen – the different cognitive languages – parameterizes the set of equivalence relations and creates the (very large) groupoid. See the mathematical appendix for a summary of standard material on groupoids.

There is a hierarchy in groupoid structures. First, there is structure *within the system having the same base and end points*, as in Ciliberti et al. Second, there is a complicated groupoid structure defined by sets of dual information sources surrounding the variation of base and end points. We do not need to know what that structure is in any detail, but can show that its existence has profound implications.

First we examine the simple case, the set of dual information sources associated with a fixed pair of beginning and end states.

Taking the serial grammar/syntax model above, we find that not all high probability meaningful paths from S_0 to S_∞ are the same. They are structured by the uncertainty of the associated dual information source, and that has a homological relation with free energy density.

Let us index possible dual information sources connecting base and end points by some set $A = \cup\alpha$. Argument by abduction from statistical physics is direct: Given metabolic energy density available at a rate M, and an allowed development time τ, let $K = 1/\kappa M \tau$ for some appropriate scaling constant κ, so that $M\tau$ is total developmental free energy. Then the probability of a particular H_α will be determined by the standard expression (e.g., Landau and Lifshitz, 2007),

$$P[H_\beta] = \frac{\exp[-H_\beta K]}{\sum_\alpha \exp[-H_\alpha K]},$$

(2.10)

where the sum may, in fact, be a complicated abstract integral.

This is just a version of the fundamental probability relation from statistical mechanics, as above. The sum in the denominator, the partition function in statistical physics, is a crucial normalizing factor that allows the definition of of $P[H_\beta]$ as a probability.

A basic requirement, then, is that the sum/integral always converges. K is the inverse product of a scaling factor, a metabolic energy density rate term, and a characteristic development time τ. The developmental energy might be raised to some power, e.g., $K = 1/(\kappa(M\tau)^b)$, suggesting the possibility of allometric scaling.

Some dual information sources will be 'richer'/smarter than others, but, conversely, must use more metabolic energy for their completion.

While we might simply impose an equivalence class structure based on equal levels of energy/source uncertainty, producing a groupoid, we can do more by now allowing both source and end points to vary, as well as by imposing energy-level equivalence. This produces a far more highly structured groupoid that we now investigate.

Equivalence classes define groupoids, by standard mechanisms (e.g., Weinstein, 1996; Brown, 1987; Golubitsky and Stewart, 2006). The basic equivalence classes – here involving both information source uncertainty level and the variation of S_0 and S_∞, will define transitive groupoids, and higher order systems can be constructed by the union of transitive groupoids, having larger alphabets that allow more complicated statements in the sense of Ash above.

Again, given an appropriately scaled, dimensionless, fixed, inverse available metabolic energy density rate and development time, so that $K = 1/\kappa M \tau$, we propose that the metabolic-energy-constrained probability of an information source representing equivalence class D_i, H_{D_i}, will again be given by the classic relation

$$P[H_{G_\alpha}] = \frac{\exp[-H_{G_\alpha} K]}{\sum_\beta \exp[-H_{G_\beta} K]},$$

where, now, we have shifted perspective, and *the sum/integral is over all possible elements of the largest available symmetry groupoid representing the equivalence class structure.* By the arguments of Ash above, compound sources, formed by the union of underlying transitive groupoids, being more complex, generally having richer alphabets, as it were, will all have higher free-energy-density-equivalents than those of the base (transitive) groupoids.

Let $Z_G \equiv \sum_\alpha \exp[-H_{G_\alpha} K]$. We now define the *Groupoid free energy* of the system, F_G, at inverse normalized metabolic energy density K, as

$$F_G[K] \equiv -\frac{1}{K} \log[Z_G[K]],$$

(2.11)

again following the standard arguments from statistical physics (again, Landau and Lifshitz, 2007, or Feynman, 2000).

2.4.1 Spontaneous symmetry breaking

The groupoid free energy permits introduction of an important idea from statistical physics.

We have expressed the probability of an information source in terms of its relation to a fixed, scaled, available (inverse) metabolic free energy density, seen as a kind of equivalent (inverse) system temperature. This gives a statistical thermodynamic path leading to definition of a 'higher' free energy construct – $F_G[K]$ – to which we now apply Landau's fundamental heuristic phase transition argument (Landau and Lifshitz 2007; Skierski et al. 1989; Pettini 2007). See, in particular, Pettini (2007) for details.

Landau's insight was that second order phase transitions were usually in the context of a significant symmetry change in the physical states of a system, with one phase being far more symmetric than the other. A symmetry is lost in the transition, a phenomenon called spontaneous symmetry breaking, and symmetry changes are inherently punctuated. The greatest possible set of symmetries in a physical system is that of the Hamiltonian describing its energy states. Usually states accessible at lower temperatures will lack the symmetries available at higher temperatures, so that the lower temperature phase is less symmetric: The randomization of higher temperatures – in this case limited by available metabolic free energy densities – ensures that higher symmetry/energy states – mixed transitive groupoid structures – will then be accessible to the system. Absent high metabolic free energy rates and densities, however, only the simplest transitive groupoid structures can be manifest. A full treatment from this perspective seems to require invocation of groupoid representations, no small matter (e.g., Buneci, 2003; Bos 2006).

Something like Pettini's (2007) Morse-Theory-based topological hypothesis can now be invoked, i.e., that changes in underlying groupoid structure are a necessary (but not sufficient) consequence of phase changes in $F_G[K]$. Necessity, but not sufficiency, is important, as it, in theory, allows mixed groupoid symmetries.

Using this formulation, the mechanisms of epigenetic catalysis are accomplished by allowing the set B_1 above to span a distribution of possible 'final' states $\mathbf{S}_\infty$. Then the groupoid arguments merely expand to permit traverse of both initial states and possible final sets, recognizing that there can now be a possible overlap in the latter, and the epigenetic effects are realized through the joint uncertainties $H(X_{G_\alpha}, Z)$, so that the epigenetic information source Z serves to direct as well the possible final states of X_{G_α}. Again, Scherrer and Jost (2007a, b) use information theory arguments to suggest something similar.

2.4.2   The groupoid atlas

The groupoid free energy inherently defines a groupoid atlas in the sense of Bak et al. (2006). Following closely Glazebrook and Wallace (2009a, b), the set of groupoids G_α comprise a groupoid atlas $\mathcal{A}$ as follows.

A family of local groupoids $(G_{\mathcal{A}})$ is defined with respective object sets $(X_{\mathcal{A}})_\alpha$, and a *coordinate system* $\Phi_{\mathcal{A}}$ of $\mathcal{A}$ equipped with a reflexive relation $\leq$. These satisfy the following conditions:

1. If $\alpha \leq \beta$ in $\Phi_{\mathcal{A}}$ then $(X_{\mathcal{A}})_\alpha \cap (X_{\mathcal{A}})_\beta$ is a union of components of $(G_{\mathcal{A}})$, that is, if $x \in (X_{\mathcal{A}})_\alpha \cap (X_{\mathcal{A}})_\beta$ and $g \in (G_{\mathcal{A}})_\alpha$ acts as $G : x \to y$, then $y \in (X_{\mathcal{A}})_\alpha \cap (X_{\mathcal{A}})_\beta$.

2. If $\alpha \leq \beta$ in $\Phi_{\mathcal{A}}$, then there is a groupoid morphism defined between the restrictions of the local groupoids to intersections

$$(G_{\mathcal{A}})_\alpha|(X_{\mathcal{A}})_\alpha \cap (X_{\mathcal{A}})_\beta \to (G_{\mathcal{A}})_\beta|(X_{\mathcal{A}})_\alpha \cap (X_{\mathcal{A}})_\beta,$$

and which is the identity morphism on objects.

Thus each of the G_α with its associated dual information source H_{G_α} constitutes a component of an atlas that incorporates the dynamicis of an interactive system by means of the intrinisic groupoid actions.

These are matters currently under very active study (e.g., del Hoyo and Minian, 2008).

2.5   Developmental holonomy

There is a more direct way to look at phase transitions in cognitive, and here culturally-driven, gene expression, adapting the topological perspectives of homotopy and holonomy directly within phenotype space.

We begin with ideas of directed homotopy.

In conventional topology one constructs equivalence classes of loops that can be continuously transformed into one another on a surface. The prospect of interest is attempt to collapse such a family of loops to a point while remaining within the surface. If this cannot be done, there is a hole. Here we are concerned, as in figure 2.1, with sets of one-way developmental trajectories, beginning with an initial phenotype $\mathbf{S_i}$, and converging on some final phenotype, here characteristic (highly dynamic) brain phenotypes labeled, respectively, $\mathbf{S_n}$ and $\mathbf{S_o}$. One might view them as, respectively, 'normal' and 'other', and the developmental pathways as representing convergence on the two different configurations. The filled triangle represents the effect of a composite external epigenetic catalyst – including the effects of culture and culturally-structured social interaction – acting at a critical developmental period represented by the initial phenotype $\mathbf{S_i}$.

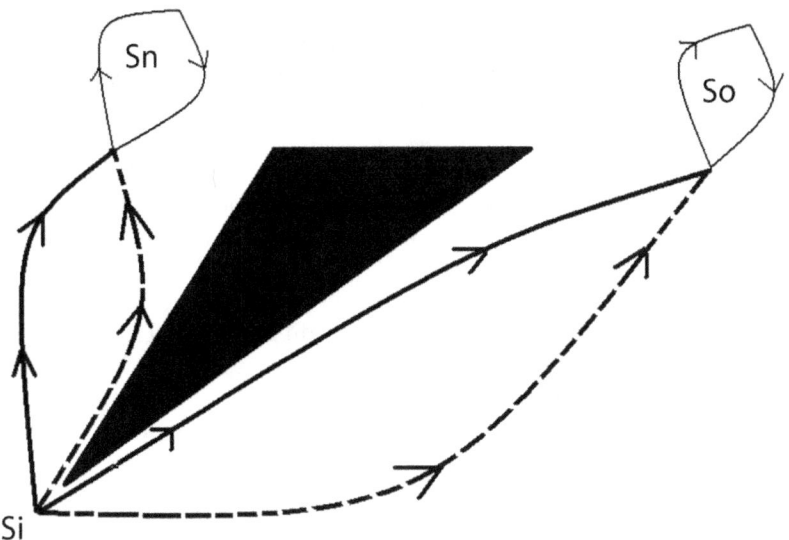

Figure 2.1: Developmental homotopy equivalence classes in phenotype space. The set on one-way paths from S_i to S_n represents an equivalence class of developmental trajectories converging on a particular phenotype, here representing a highly dynamic normal mind structure. In the presence of a noxious external epigenetic catalyst, developmental trajectories can converge on an abnormal mind structure, represented by the dynamic phenotype S_o.

We assume phenotype space to be directly measurable and to have a simple 'natural' metric defining the difference between developmental paths.

Developmental paths continuously transformable into each other without crossing the filled triangle define equivalence classes characteristic of different information sources dual to cognitive gene expression, as above.

Given a metric on phenotype space, and given equivalence classes of developmental trajectories having more than one path each, we can *pair one-way developmental trajectories* to make loop structures. In figure 2.1 the solid and dotted lines above and below the filled triangle can be pasted together to make loops characteristic of the different developmental equivalence classes. Although figure 2.1 is represented as topologically flat, there is no inherent reason for the phenotype manifold itself to be flat. The existence of a metric in phenotype space permits determining the degree of curvature, using standard methods. Using the metric definition it is possible to *parallel transport* a tangent vector starting at point s around the loop, and to measure the angle between the initial and final vectors, as indicated. A central result from elementary metric geometry is that the angle α will be given by the integral of the curvature tensor of the metric over the interior of the loop (Frankel, 2006, Section 9.6).

The *holonomy group* is defined as follows (Helgason, 1962):

If s is a point on a manifold M having a metric, then the holonomy group of M is the group of all linear transformations of the tangent space M_s obtained by parallel translation along closed curves starting at s.

For figure 2.1 the *phenotype holonomy groupoid* is the disjoint union of the different holonomy groups corresponding to the different branches separated by 'developmental shadows' induced by epigenetic information sources acting as developmental catalysts.

The relation between the phenotype groupoid as defined here and the phase transitions in $F_D[K]$ as defined above is an open question, and is a central focus of ongoing work.

2.6   The manifold of dual information sources

2.6.1   Basic structure

Glazebrook and Wallace (2009a) examined holonomy groupoid phase transition arguments for networks of interacting information sources dual to cognitive phenomena. A more elementary form of this arises directly through extending holonomy groupoid arguments to a manifold of different information source dual to cognitive phenomena as follows.

Different cognitive phenomena will have different dual information sources, and we are interested in the local properties of the system near a particular reference state. We impose a topology on the system, so that, near a particular 'language' A, dual to an underlying cognitive process, there is an open set U of closely similar languages $\hat{A}$, such that $A, \hat{A} \subset U$. It may be necessary to coarse-grain the system's responses to define these information sources. The problem is to proceed in such a way as to preserve the underlying essential topology, while eliminating 'high frequency noise'. The formal tools for this can be found elsewhere, e.g., in Chapter 8 of Burago et al. (2001).

Since the information sources dual to the cognitive processes are similar, for all pairs of languages $A, \hat{A}$ in U, it is possible to:

[1] Create an embedding alphabet which includes all symbols allowed to both of them.

[2] Define an information-theoretic distortion measure in that extended, joint alphabet between any high probability (grammatical and syntactical) paths in A and $\hat{A}$, which we write as $d(Ax, \hat{A}x)$ (Cover and Thomas, 1991). Note that these languages do not interact, in this approximation.

[3] Define a metric on U, for example,

$$\mathcal{M}(A, \hat{A}) = |\lim \frac{\int_{A, \hat{A}} d(Ax, \hat{A}x)}{\int_{A, A} d(Ax, A\hat{x})} - 1|,$$

(2.12)

using an appropriate integration limit argument over the high probability paths. Note that the integration in the denominator is over different paths within A itself, while in the numerator it is between different paths in A and $\hat{A}$. Other metric constructions on U seem possible. Structures weaker than a conventional metric would be of more general utility, but the mathematical complications are formidable.

Note that these conditions can be used to define equivalence classes of *languages* dual to cognitive processes, where previously we defined equivalence classes of *states* that could be linked by high probability, grammatical and syntactical paths connecting two phenotypes. This led to the characterization of different information sources. Here we construct an entity, formally a topological manifold, *that is an equivalence class of information sources*. This is, provided $\mathcal{M}$ is a conventional metric, a classic differentiable manifold. The set of such equivalence classes generates the *dynamical groupoid*, and questions arise regarding mechanisms, internal or external, which can break that groupoid symmetry.

Since H and $\mathcal{M}$ are both scalars, a 'covariant' derivative can be defined directly as

$$dH/d\mathcal{M} = \lim_{\hat{A} \to A} \frac{H(A) - H(\hat{A})}{\mathcal{M}(A, \hat{A})},$$

(2.13)

where $H(A)$ is the source uncertainty of language A.

Suppose the system to be set in some reference configuration A_0.

To obtain the unperturbed dynamics of that state, impose a Legendre transform using this derivative, defining another scalar

$$S \equiv H - \mathcal{M}dH/d\mathcal{M}.$$

(2.14)

The simplest possible generalized Onsager relation – here seen as an empirical, fitted, equation like a regression model – is

$$d\mathcal{M}/dt = LdS/d\mathcal{M},$$

(2.15)

where t is the time and $dS/d\mathcal{M}$ represents an analog to the thermodynamic force in a chemical system. This is seen as acting on the reference state A_0.

Explicit parameterization of $\mathcal{M}$ introduces standard – and quite considerable – notational complications (Burago et al., 2001): Imposing a metric for different cognitive dual languages parameterized by $\mathbf{K}$ leads to Riemannian, or even Finsler, geometries, including the usual geodesics (e.g., Wallace and Fullilove, 2008; Glazebrook and Wallace, 2009a, b).

The dynamics, as we have presented them so far, have been noiseless. The simplest generalized Onsager relation in the presence of noise might be rewritten as

$$d\mathcal{M}/dt = LdS/d\mathcal{M} + \sigma W(t),$$

where σ is a constant and $W(t)$ represents white noise. Again, S is seen as a function of the parameter $\mathcal{M}$. This leads directly to a family of classic stochastic differential equations of the form

$$d\mathcal{M}_t = L(t, \mathcal{M})dt + \sigma(t, \mathcal{M})dB_t,$$

(2.16)

where L and σ are appropriately regular functions of t and $\mathcal{M}$, and dB_t represents the noise structure, characterized by its quadratic variation.

In the sense of Emery (1989), this leads into complicated realms of stochastic differential geometry and related topics.

See the Mathematical Appendix for an introduction to stochastic differential equations from a Martingale perspective.

The natural generalization is to a system of developmental processes that interact via mutual information crosstalk, as described by Wallace and Wallace (2009).

2.6.2 'Coevolutionary' development

The model can be applied to multiple interacting information sources representing simultaneous gene expression processes. This is, in a broad sense, a 'coevolutionary' phenomenon in that the development of one process may affect that of others.

Most generally we assume that different cognitive developmental subprocesses of gene expression characterized by information sources H_m interact through chemical or other signals and assume that *different processes become each other's principal environments*, a broadly coevolutionary phenomenon.

We write

$$H_m = H_m(K_1...K_s, ...H_j...),$$

(2.17)

where the K_s represent other relevant parameters and $j \neq m$.

The dynamics of such a system is driven by a recursive network of stochastic differential equations, similar to those used to study many other highly parallel dynamic structures (Zhu et al., 2007).

Letting the K_j and H_m all be represented as parameters Q_j, (with the caveat that H_m not depend on itself), one can define, according to the generalized Onsager development of Wallace and Wallace (2009),

$$S^m = H_m - \sum_i Q_i \partial H_m / \partial Q_i$$

to obtain a complicated recursive system of phenomenological 'Onsager relations' stochastic differential equations,

$$dQ_t^j = \sum_i [L_{j,i}(t, ...\partial S^m / \partial Q^i ...)dt + \sigma_{j,i}(t, ...\partial S^m / \partial Q^i ...)dB_t^i],$$

(2.18)

where, again for notational simplicity only, we have expressed both the H_j and the external K's in terms of the same symbols Q_j.

m ranges over the H_m and we could allow different kinds of 'noise' dB_t^i, having particular forms of quadratic variation that may, in fact, represent a projection of environmental factors under something like a rate distortion manifold (Glazebrook and Wallace, 2009).

It is important to realize that, for this formulation, one does not necessarily have the equivalent of 'Onsager's fourth law' of thermodynamics, i.e., the symmetry relation $L_{i,j} = L_{j,i}$. This is because such a symmetry, at base, is a statement of local time reversal invariance (e.g., de Groot and Mazur, 1984, pp. 35-41). But information sources are notoriously one-way in time, for example someone speaking or writing in English is much more likely to utter the five-character string " the " than its reverse. Or, to put it another way, long palindromes, such as "Able was I ere I saw Elba", or "A man, a plan, a canal: Panama", are quite rare, and always relatively short, while information theory is based on asymptotic limit theorems most often involving very long strings of symbols.

As usual, for a system of equations like (2.18), there will be multiple quasi-stable points, representing a class of generalized resilience modes accessible via holonomy punctuation.

There are, indeed, many possible patterns:

1. Setting equation (2.18) equal to zero and solving for stationary points gives attractor states since the noise terms preclude unstable equilibria.

2. This system may, however, converge to limit cycle or 'strange attractors' that are very highly dynamic.

3. What is converged to in both cases is not a simple state or limit cycle of states. Rather it is an equivalence class, or set of them, of generalized language information sources coupled by mutual interaction through crosstalk. Thus 'stability' in this extended model represents particular patterns of ongoing dynamics rather than some identifiable 'state', although such dynamics may be indexed by a 'stable' set of phenotypes.

Here we become enmeshed in a system of highly recursive phenomenological stochastic differential equations, but at a deeper level than the standard stochastic chemical reaction model (Zhu et al., 2007), and in a dynamic rather than static manner: the objects of this system are equivalence classes of information sources and their crosstalk, rather than simple final states of a chemical system.

We have defined a groupoid for the system based on a particular set of equivalence classes of information sources dual to cognitive processes. That groupoid parsimoniously characterizes the available dynamical manifolds, and breaking of the groupoid symmetry by epigenetic crosstalk creates more complex objects of considerable interest. This leads to the possibility, indeed, the necessity of epigenetic *Deus ex Machina* mechanisms – analogous to programming, stochastic resonance, etc. – to force transitions between the different possible modes within and across dynamic manifolds. In one model the external 'programmer' creates the manifold structure, and the system hunts within that structure for the 'solution' to the problem according to equivalence classes of paths on the manifold. Noise, as with random mutation in evolutionary algorithms, precludes simple unstable equilibria, but not other possible structures.

Equivalence classes of *states* gave dual information sources. Equivalence classes of *information sources* give different characteristic dynamical manifolds. Equivalence classes of one-way developmental *paths* produce different directed homotopy topologies characterizing those dynamical manifolds. This introduces the possibility of having different quasi-stable modes *within* individual manifolds, and leads to ideas of holonomy and the holonomy groupoid of the set of quasi-stable developmental modes.

2.7 Rate distortion models

2.7.1 The Rate Distortion Theorem

The interaction between cognitive structures can be restated from a highly formal, but more restricted, Rate Distortion Theorem perspective. Suppose a sequence of signals is generated by an information source dual to a cognitive process, Y having output $y^n = y_1, y_2, \ldots$. This is 'digitized' in terms of the observed behavior of the system with which it communicates, say a sequence of observed behaviors $b^n = b_1, b_2, \ldots$. Often the b_i will happen in a characteristic 'real time' τ. Assume each b^n is then deterministically retranslated back into a reproduction of the original biological signal,

$$b^n \to \hat{y}^n = \hat{y}_1, \hat{y}_2, \ldots.$$

Define a distortion measure $d(y, \hat{y})$ that compares the original to the retranslated path. Many such measures are possible. The Hamming distortion, for example, is

$$d(y, \hat{y}) = 1, y \neq \hat{y}$$

$$d(y, \hat{y}) = 0, y = \hat{y}$$

For continuous variates the squared error distortion is

$$d(y, \hat{y}) = (y - \hat{y})^2.$$

The distortion between *paths* y^n and $\hat{y}^n$ is defined as

$$d(y^n, \hat{y}^n) \equiv \frac{1}{n} \sum_{j=1}^{n} d(y_j, \hat{y}_j).$$

A remarkable fact of the Rate Distortion Theorem is that *the basic result is independent of the exact distortion measure chosen* (Cover and Thomas, 1991; Dembo and Zeitouni, 1998).

Suppose that with each path y^n and b^n-path retranslation into the y-language, denoted $\hat{y}^n$, there are associated individual, joint, and conditional probability distributions

$$p(y^n), p(\hat{y}^n), p(y^n, \hat{y}^n), p(y^n | \hat{y}^n).$$

The average distortion is defined as

$$D \equiv \sum_{y^n} p(y^n) d(y^n, \hat{y}^n).$$
(2.19)

It is possible, using the distributions given above, to define the information transmitted from the Y to the $\hat{Y}$ process using the Shannon source uncertainty of the strings:

$$I(Y, \hat{Y}) \equiv H(Y) - H(Y|\hat{Y}) = H(Y) + H(\hat{Y}) - H(Y, \hat{Y}),$$

(2.20)

where $H(..., ...)$ is the joint and $H(...|...)$ the conditional uncertainty (Cover and Thomas, 1991; Ash, 1990).

If there is no uncertainty in Y given the retranslation $\hat{Y}$, then no information is lost, and the systems are in perfect synchrony.

In general, of course, this will not be true.

The *rate distortion function* $R(D)$ for a source Y with a distortion measure $d(y, \hat{y})$ is defined as

$$R(D) = \min_{p(y,\hat{y}); \sum_{(y,\hat{y})} p(y)p(y|\hat{y})d(y,\hat{y}) \leq D} I(Y,\hat{Y}).$$

(2.21)

The minimization is over all conditional distributions $p(y|\hat{y})$ for which the joint distribution $p(y,\hat{y}) = p(y)p(y|\hat{y})$ satisfies the average distortion constraint (i.e., average distortion $\leq D$).

The *Rate Distortion Theorem* states that $R(D)$ is the minimum necessary rate of information transmission – minimum channel capacity – that ensures the communication between the modules does not exceed average distortion D. Thus $R(D)$ defines a minimum necessary channel capacity. Cover and Thomas (1991) or Dembo and Zeitouni (1998) provide details. The rate distortion function has been calculated for a number of systems.

There is an absolutely central fact characterizing the rate distortion function: Cover and Thomas (1991, Lemma 13.4.1) show that $R(D)$ *is necessarily a decreasing convex function of* D for any reasonable definition of distortion.

That is, $R(D)$ *is always* a reverse J-shaped curve. This will prove crucial for the overall argument. Indeed, convexity is an exceedingly powerful mathematical condition, and permits deep inference (e.g., Rockafellar, 1970). Ellis (1985, Ch. VI) applies convexity theory to conventional statistical mechanics.

For a Gaussian channel having noise with zero mean and variance σ^2 (Cover and Thomas, 1991),

$$R(D) = 1/2 \log[\sigma^2/D], 0 \leq D \leq \sigma^2,$$

$$R(D) = 0, D > \sigma^2.$$

(2.22)

Recall, now, the relation between information source uncertainty and channel capacity (e.g., Ash, 1990):

$$H[\mathbf{X}] \leq C,$$

(2.23)

where H is the uncertainty of the source X and C the channel capacity, defined according to the relation (Ash, 1990)

$$C \equiv \max_{P(X)} I(X|Y),$$

(2.24)

where $P(X)$ is chosen so as to maximize the rate of information transmission along a channel Y.

Finally, recall the analogous definition of the rate distortion function from equation (2.21), again an extremum over a probability distribution.

2.7.2   Rate Distortion Dynamics

$R(D)$ defines the minimum channel capacity necessary for the system to have average distortion less than or equal D, placing a limit on information source uncertainty. Thus, we suggest distortion measures can drive information system dynamics. That is, the rate distortion function also has a homological relation to free energy density, similar to the relation between free energy density and information source uncertainty.

We are led to propose, as a heuristic, that the dynamics of cognitive modules interacting in a characteristic 'real time' τ will be constrained by the system as described in terms of a parameterized rate distortion function. To do this, take R as parameterized, not only by the distortion D, but by some vector of variates $\mathbf{Q} = (Q_1, ..., Q_k)$, for which the first component is the average distortion. The assumed dynamics are, as in Wallace and Wallace, (2008), then driven by gradients in the rate distortion disorder defined as

$$S_R \equiv R(\mathbf{Q}) - \sum_{i=1}^{k} Q_i \partial R / \partial Q_i.$$

(2.25)

This leads to the deterministic and stochastic systems of equations analogous to the Onsager relations of nonequilibrium thermodynamics:

$$dQ_j/dt = \sum_i L_{j,i} \partial S_R / \partial Q_i$$

(2.26)

and

$$dQ_t^j = L^j(Q_1, ..., Q_k, t)dt + \sum_i \sigma^{j,i}(Q_1, ..., Q_k, t)dB_t^i,$$

(2.27)

where the dB_t^i represent added, often highly structured, stochastic 'noise' whose properties are characterized by the quadratic variation (e.g., Protter, 1995).

Even for this simplified structure, it is not clear under what circumstances 'Onsager reciprocal relations' are possible. Since average distortion is a scalar, however, some systems may indeed display the kind of time reversal invariance required for those symmetries.

A central focus of this chapter, however, is to generalize these equations in the face of richer structures, for example interactions between cognitive modules that may not be time-reversible, the existence of characteristic time constants within nested processes, and the influence of an embedding source of free energy.

For a simple Gaussian channel with noise having zero mean and variance σ^2,

$$S_R(D) = R(D) - DdR(D)/dD = 1/2 \log(\sigma^2/D) + 1/2.$$

(2.28)

The simplest possible Onsager relation becomes

$$dD/dt = -\mu dS_R/dD = \frac{\mu}{2D},$$

(2.29)

where $-dS_R/dD$ represents the force of an entropic wind, a kind of internal dissipation inevitably driving the real-time, system of interacting (cognitive) information sources toward greater distortion.

This has the solution

$$D = \sqrt{\mu t},$$

(2.30)

so that the average distortion increases monotonically with time, for this model.

An important observation is that *similar results must necessarily apply to any of the reverse-J-shaped relations that inevitably characterize* $R(D)$, since the rate distortion function is necessarily a convex decreasing function of the average distortion D, whatever distortion measure is chosen. Again, see Cover and Thomas (1991, Lemma 13.4.1) for details.

The explicit implication is that a system of cognitive modules interacting in real time will inevitably be subject to a relentless entropic force, requiring a constant free energy expenditure for maintenance of some fixed average distortion in the communication between them: The distortion in the communication between two interacting modules will, without free energy input, have time dependence

$$D = f(t),$$

(2.31)

with $f(t)$ monotonic increasing in t.

This necessarily leads to the punctuated failure of the system.

Note that equation (2.30) is similar to classical Brownian motion as treated by Einstein: Let $p(x,t)dx$ be the probability a particle located at the origin at time zero and undergoing Brownian motion is found at locations $x \to x + dx$ at time t. Then, p satisfies the diffusion equation $\partial p(x,t)/\partial t = \mu \partial^2 p(x,t)/\partial x^2$. Einstein's solution is that

$$p(x,t) = \frac{1}{\sqrt{4\pi\mu t}} \exp[-x^2/4\mu t].$$

It is easy to show that the standard deviation of the particle position increases in proportion to $\sqrt{\mu t}$, just as above.

Some comment is appropriate. Following Chung and Williams (1990), a process $B = B_t, t \in \mathcal{R}_+$ is called a Brownian motion in $\mathcal{R}_+$ iff:

[1] for $0 \leq s < t < \infty, B_t - B_s$ is a normally distributed random variate with mean zero and variance $|t - s|$.

[2] for $0 \leq t_0 < t_1 < ... < t_k < \infty$,

$$\{B_{t_0}; B_{t_j} - B_{t_{j-1}}, j = 1, ..., k\}$$

is a set of independent random variates.

An information source, of course, generates a *highly correlated sequence* that grossly violates these simple assumptions. What we have shown is that the *distortion* in the communication between interacting cognitive modules, under appropriate empirical Onsager relations, can behave as if it were undergoing Brownian motion.

This is a simple, but far from trivial, result.

Prandolini and Moody (1995) have, in fact, observed something much like this in the time base error of recorded signals. Wow and flutter are the instantaneous speed error between recording and reproduction epochs. The time base error (TBE) in the reproduced signal is a function of the wow and flutter. They show, empirically, that the nonperiodic TBE is a *fractional Brownian motion*. The implication is that the nonperiodic flutter is fractional Gaussian, and thus what they call a 'blind' TBE system is impractical for the design of a TBE compensation system.

Normalized fractional Brownian motion on $(0,t), t \in \mathcal{R}_+$ is a continuous time Gaussian process starting at zero, with mean zero, and having the covariance function (Beran, 1994)

$$E[B^H(t)B^H(s)] = (1/2)[|t|^{2H} + |s|^{2H} - |t - s|^{2H}].$$

If $H = 1/2$ the process is a regular Brownian motion. Otherwise, for $H > 1/2$, the increments are positively correlated, and for $H < 1/2$, negatively correlated.

We will explore this kind of relation in more detail below.

2.7.3   Rate distortion coevolutionary dynamics

A simplified version of equation (2.18) can be constructue using the rate distortion functions for mutual crosstalk between a set of interacting cognitive modules, using the homology of the rate distortion function itself with free energy, as driven by the inherent convexity of the Rate Distortion Function $R(D)$. That convexity is, in fact, why we invoke the Rate Distortion Function.

Given different cognitive processes $1...s$, the quantities of special interest thus become the mutual rate distortion functions $R_{i,j}$ characterizing communication (and the distortion $D_{i,j}$) between them, while the essential parameters remain the characteristic time constants of each process, $\tau_j, j = 1...s$, and an overall, embedding, available free energy density, F.

Taking the Q^α to run over all the relevant parameters and mutual rate distortion functions (including distortion measures $D_{i,j}$), equation (14) becomes

$$S_R^{i,j} \equiv R_{i,j} - \sum_k Q^k \partial R_{i,j}/\partial Q^k.$$

(2.32)

Equation (2.18) accordingly becomes

$$dQ_t^\alpha = \sum_{\beta=(i,j)} [L_\beta(t,...\partial S_R^\beta/\partial Q^\alpha...)dt + \sigma_\beta(t,...\partial S_R^\beta/\partial Q^\alpha...)dB_t^\beta],$$

(2.33)

and this generalizes the treatment in terms of crosstalk, its distortion, the inherent time constants of the different cognitive modules, and the overall available free energy density.

This is a very complicated structure indeed, but its general dynamical behaviors will obviously be analogous to those described just above. For example, setting equation (2.33) to zero gives the 'coevolutionary stable states' of a system of interacting cognitive modules. Again, limit cycles and strange attractors seem possible as well. And again, what is converged to is a dynamic behavior pattern, not some fixed 'state'. And again, such a system will display highly punctuated dynamics almost exactly akin to resilience domain shifts in ecosystems (Holling, 1973, 1992; Gunderson, 2000). Indeed, the formalism seems directly applicable to ecosystem studies.

And again, because these are highly self-dynamic cognitive phenomena and not simple crystals or other physical objects, it may not often be possible to invoke time reversal invariance to give Onsager-like reciprocal symmetries to equation (2.33).

2.7.4 An example

First, assume a fixed embedding communication free energy density of F, representing the richness of incoming information from the interacting cognitive modules. The simplest generalization of equation (2.29), for a Gaussian channel, becomes

$$dD/dt = \mu/2D - \alpha F, \alpha > 0,$$

(2.34)

representing the distortion in the communication between two interacting cognitive processes.

This has the equilibrium solution for the distortion,

$$D_{equlib} = \frac{\mu}{2\alpha F}.$$

(2.35)

In contrast to equation (2.29), where, in the absence of some free energy/information input, the distortion grows as the square root of the elapsed time, here there is a finite, equilibrium, average distortion, that is inversely proportional to the available environmental or informational free energy, that the interacting systems can use to direct their actions.

The obvious generalization is

$$D_{equilib} = \frac{1}{g(F)},$$

(2.36)

where $g(F)$ is monotonic increasing in F.

Introducing a characteristic response time variable τ, so that

$$dD/dt = \frac{\mu}{2D} - g(F)h(\tau),$$

(2.37)

where $h(\tau)$ is also monotonic increasing, leads to

$$D_{equilib} = \frac{\mu}{2g(F)h(\tau)}.$$

(2.38)

Thus, for this particular phenomenological Onsager model, at a fixed rate of available information free energy, increasing allowable response time decreases average distortion in the interaction between the cognitive subsystems.

This is, in fact, a classic result across a broad spectrum of engineering applications.

If we now allow feedback, so that the system actively seeks information in proportion to the distortion between intent and impact, then the empirical Onsager relation for a Gaussian channel becomes

$$dD/dt = \frac{\mu}{2D} - g(F)h(\tau)D,$$

(2.39)

and

$$D_{equilib} = \sqrt{\frac{\mu}{2g(F)h(\tau)}},$$

(2.40)

significantly smaller than (2.38).

This is, in fact, precisely the classic result for Brownian motion in a harmonic central field (Wang and Uhlenbeck, 1945, eq. 54), restated in terms of average distortion.

A mixed strategy,

$$dD/dt = \frac{\mu}{2D} - g(F)h(\tau)[1 + \alpha D],$$

(2.41)

leading to the quadratic

$$2Dg(F)h(\tau)(1 + \alpha D) - \mu = 0,$$

(2.42)

has a single equilibrium solution

$$D_{equilib} = \frac{-g(F)h(\tau) + \sqrt{g(F)^2 h(\tau)^2 + 2g(F)h(\tau)\alpha\mu}}{2g(F)h(\tau)\alpha},$$

(2.43)

since D must be greater than zero and real.

Stochastic generalizations – the diffusion of distortion as it were – involving probabilistic fuzz about deterministic equilibria or dynamic paths, seem direct.

2.8 Expanding the theory

We have, in the context of the tunable epigenetic catalysis of Wallace and Wallace (2009), developed three separate phase transition/branching models of cognitive gene expression based on groupoid structures that may be applied to the development of the human mind and its dysfunctions, as known to be particularly influenced by embedding culture. The first used Landau's spontaneous symmetry breaking to explore phase transitions in a groupoid free energy $F_D[K]$. The second examined a holonomy groupoid in phenotype space generated by disjoint developmental homotopy equivalence classes, and 'loops' constructed by pairing one-way development paths. The third introduced a metric on a manifold of different information sources dual to cognitive gene expression, leading to a more conventional picture of parallel transport around a loop leading to holonomy. The dynamical groupoid of Wallace and Fullilove (2008, Sec. 3.8) is seen as involving a disjoint union across underlying manifolds that produces a holonomy groupoid in a natural manner.

There are a number of outstanding mathematical questions.

The first is the relation between the Landau formalism and the structures of phenotype space S and those of the associated manifold of dual information sources, the manifold M having metric $\mathcal{M}$. How does epigenetic catalysis in M-space imposes structure on S-space? How is this related to spontaneous symmetry breaking?

What would a stochastic version of the theory, in the sense of Emery (1989), look like? It is quite possible, using appropriate averages of the stochastic differential equations that arise naturally, to define parallel transport, holonomy, and the like for these structures. In particular a stochastic extension of the results of the first question would seem both fairly direct and interesting from a real-world perspective, as development is always 'noisy'.

The construction of loops from directed homotopy arcs in figure 2.1 is complicated by the necessity of imposing a consistent piecewise patching rule for parallel translation at the end of each arc, say from $\mathbf{S}_i$ to $\mathbf{S}_n$. This can probably be done by some standard fiat, but the details will likely be messy.

On another matter, we have imposed metrics on S and M space, making possible a fairly standard manifold analysis of complex cognitive processes of gene expression and development. While this is no small thing, an important 'natural' generalization, given the ubiquity of groupoids across our formalism, would be to a *groupoid atlas*, in the sense of Bak et al., (2006) and Glazebrook and Wallace (2009b, Section 7.4). The groupoid atlas permits a weaker structure compared with that of a conventional manifold since no condition of compatibility between arbitrary overlaps of the patches is necessary. It is possible that the groupoid atlas will become, to complicated problems in biological cognitive process, something of what the Riemannian manifold has been to physics. The groupoid atlas, unlike the Riemannian manifold, is quite new and under active study.

With regard to questions of 'smoothness', we are assuming that the cognitive landscape of gene expression is sufficiently rich that discrete paths can be well approximated as continuous where necessary, the usual physicist's hack.

Finally, sections 2.5 and 2.6 are based on existence of more-or-less conventional metrics, and this may not be a good approximation to many real systems. Extending topological phase transition theory to 'weaker' topologies, e.g., Finsler geometries and the like, is not a trivial task.

2.9 Discussion

Culturally structured psychosocial stress, and similar noxious exposures, can write distorted images of themselves onto human ontology – both child growth, and, if sufficiently powerful, adult development as well – by a variety of mechanisms, initiating a punctuated trajectory to characteristic forms of comorbid mind/body dysfunction. This occurs in a manner recognizably analogous to resilience domain shifts affecting stressed ecosystems (e.g., Wallace, 2008; Holling, 1973; Gunderson, 2000). Consequently, like ecosystem restoration, reversal or palliation may often be exceedingly difficult once a generalized domain shift has taken place. Thus a public health approach to the prevention of mental disorders may be paramount: rather than seeking to understand why half a population does not respond to the LD50 of a teratogenic environmental exposure, one seeks policies and social reforms that limit the exposure.

Both socio-cultural and epigenetic environmental influences – like gene methylation – are heritable, in addition to genetic mechanisms. The missing heritability of complex diseases that Manolio et al. (2009) seek to find in more and better gene studies is most likely dispersed within the 'dark matter' of these two other systems of heritage that together constitute the larger, and likely highly synergistic, regulatory machinery for gene expression. More and more purely genetic studies would, under such circumstances, be akin to using increasingly powerful microscopes to look for cosmic membranes of strewn galaxies.

A crucial matter is the conversion of the probability models we present here into statistical tools suitable for analyzing real data. Some work in this direction has been done in Section 2.7, but the problem involves not just programming such models for use, but identifying appropriate real-world problems, assembling available data sets, transforming the data as needed for the models, and actually applying the statistical models. Indeed, the environmental health literature contains numerous examples of developmental deviations due to either chemical exposures or interaction between chemical and socioeconomic exposures, and these could serve as sources of data for direct analysis (e.g., Needleman et al., 1996; Fullilove, 2004; Dietrich et al., 2001; Miranda et al., 2007; Glass et al., 2009; Jacobson and Jacobson, 2003; Shankardass et al., 2009; Clougherty et al., 2007; Ben Jonathan et al., 2009; Karp et al., 2005; Sarlio-Lahteenkorva and Lahelma, 2001; Wallace and Wallace, 2005; Wallace, Wallace and Rauh, 2003). Thus, quite a number of data sets exist in the environmental health and socioeconomic epidemiological literature that could be subjected to meta-analysis and other review for model verification and fitting. Our topological models, when converted to statistical tools for data analysis, hold great potential for understanding developmental trajectories and interfering factors (teratogens) through the life course. Sets of cross cultural variants of these data focusing specifically on mental disorders, would be needed to address the particular concerns of this paper.

Nonetheless, what we have done is of no small interest for understanding the ontology of the human mind and its pathologies. West-Eberhard (2003, 2005) argues that any new input, whether it comes from the genome, like a mutation, or from the external environment, like a temperature change, a pathogen, or a parental opinion, has a developmental effect only if the preexisting phenotype is responsive to it. A new input causes a reorganization of the phenotype, or 'developmental recombination'. In developmental recombination, phenotypic traits are expressed in new or distinctive combinations during ontogeny, or undergo correlated quantitative change in dimensions. Developmental recombination can result in evolutionary divergence at all levels of organization.

According to West-Eberhard, individual development can be visualized as a series of branching pathways. Each branch point is a developmental decision, or switch point, governed by some regulatory apparatus, and each switch point defines a modular trait. Developmental recombination implies the origin or deletion of a branch and a new or lost modular trait. The novel regulatory response and the novel trait originate simultaneously. Their origins are, in fact, inseparable events: There cannot, West-Eberhard concludes, be a change in the phenotype, a novel phenotypic state, without an altered developmental pathway.

Our analysis provides a new formal picture of this process as it applies to the human mind: The normal branching of developmental trajectories, and the disruptive impacts of teratogeneic events of various kinds, can be described in terms of a growing sequence of holonomy groupoids, each associated with a set of dual information sources representing patterns of cognitive gene expression catalyzed by epigenetic information sources that, for humans, must include culture and culturally-modulated social interaction as well as more direct mechanisms like gene methylation. This is a novel way of looking at human mental development and its disorders that may prove to be of some use. The most important innovation of this work, however, seems to be the natural incorporation of embedding culture as an essential component of the epigenetic regulation of human mental development, and in the effects of environment on the expression of mental disorders, bringing what is perhaps the central reality of human biology into the center of contemporary biological psychiatry.

In sum, we have outlined a broad class of probability models of gene-culture-environment interaction, and outlined some statistical models based on them, that might help current studies of gene-environment interaction in American psychiatry avoid Heine's (2001) trap of developing an understanding of the self, and its disorders, that is peculiar in the context of the world's cultures.

2.10 References

Arnett, J., 2008, The neglected 95 %, *The American Psychologist*, 63:602-614.

Ash, R., 1990, *Information Theory*, Dover Publications, New York.

Atlan, H., and I. Cohen, 1998, Immune information, self-organization, and meaning, *International Immunology*, 10:711-717.

Atmanspacher, H., 2006, Toward an information theoretical implementation of contextual conditions for consciousness, *Acta Biotheoretica*, 54:157-160.

Baars, B., 1989, *A Cognitive Theory of Consciousness*, Cambridge University Press, New York.

Backdahl, L., A. Bushell, and S. Beck, 2009, Inflammatory sig-

nalling as mediator of epigenetic modulation in tissue-specific chronic inflammation, *The International Journal of Biochemistry and Cell Biology*,
doi:10.1016/j.biocel.2008.08.023.

Bak, A., R. Brown, G. Minian, and T. Porter, 2006, Global actions, groupoid atlases and related topics, *Journal of Homotopy and Related Structures*, 1:1-54.

Bebbington, P., 1993, Transcultural aspects of affective disorders, *International Review of Psychiatry*, 5:145-156.

Ben-Jonathan, N., E. Hugo, T. Brandenbourg, 2009, Effects of bisphenol A on adipokine release from human adipose tissue: implications for the metabolic syndrome, *Molecular Cell Endocrinology*, 304:49-54.

Bennett, C., 1988, Logical depth and physical complexity. In *The Universal Turing Machine: A Half-Century Survey*, R. Herkin (ed.), pp. 227-257, Oxford University Press.

Beran, J., 1994, *Statistics for Long-Memory Processes*, Chapman and Hall, New York.

Bos, R., 2007, Continuous representations of groupoids, arXiv:math/0612639.

Bossdorf, O., C. Richards, and M. Pigliucci, 2008, Epigenetics for ecologists, *Ecology Letters*, 11:106-115.

Brown, G., T. Harris, and J. Peto, 1973, Life events and psychiatric disorders, II: nature of causal link, *Psychological Medicine*, 3:159-176.

Brown, R., 1987, From groups to groupoids: a brief survey, *Bulletin of the London Mathematical Society*, 19:113-134.

Buneci, M., 2003, *Representare de Groupoizi*, Editura Mirton, Timisoara.

Burago, D., Y. Burago, and S. Ivanov, 2001, *A Course in Metric Geometry*, Graduate Studies in Mathematics 33, American Mathematical Society.

Burns, J., D. Job, M. Bastin, H. Whalley, T. Macgillivary, E. Johnstone, and S. Lawrie, 2003, Structural disconnectivity in schizophrenia: a diffusion tensor magnetic resonance imaging study, *British Journal of Psychiatry*, 182:439-443.

Cannas Da Silva, A., and A. Weinstein, 1999, *Geometric Models for Noncommutative Algebras*, American Mathematical Society, RI.

Caspi, A., et al., 2002, Role of genotype in the cycle of violence in maltreated children, *Science*, 297:851-854.

Caspi, A., K. Sugden, T. Moffitt, et al., 2003, Influence of life stress on depression : moderation by a polymorphism in the 5-HTT gene, *Science*, 301:386-389.

Caspi, A., and T. Moffitt, 2006, Gene-environment interactions in psychiatry: joining forces with neuroscience, *Nature Reviews Neuroscience* 7:583-590.

Chung, K., and R. Williams, 1990, *Introduction to Stochastic Integration*, Second Edition, Birkhauser, Boston, MA.

Ciliberti, S., O. Martin, and A. Wagner, 2007a, Robustness can evolve gradually in complex regulatory networks with varying topology, *PLoS Computational Biology*, 3(2):e15.

Ciliberti, S., O. Martin, and A. Wagner, 2007b, Innovation and robustness in complex regulatory gene networks, *Proceeding of the National Academy of Sciences*, 104:13591-13596.

Clougherty, J., J. Levy, L. Kubzansky, P. Ryan, S. Suglia, M. Canner, and R. Wright, 2007, Synergistic effects of traffic-related air pollution and exposure to violence on urban asthma etiology, *Environmental Health Perspectives*, 115:1140-1146.

Cohen, I., 2006, Immune system computation and the immunological homunculus. In Nierstrasz, O., J. Whittle, D. Harel, and G. Reggio (eds.), MoDELS 2006, LNCS, vol. 4199, pp. 499-512, Springer, Heidelberg.

Cohen, I., and D. Harel, 2007, Explaining a complex living system: dynamics, multi-scaling, and emergence. *Journal of the Royal Society: Interface*, 4:175-182.

Coplan, J., 2005, personal communication.

Courchesne, E., and K. Pierce, 2005, *Current Opinion in Neurobiology*, 15:225-230.

Cover, T., and J. Thomas, 1991, *Elements of Information Theory*, John Wiley and Sons, New York.

deGroot S., and R. Mazur, 1984, *Non-Equilibrium Thermodynamics*, Dover Publications, New York.

Dehaene, S., and L. Naccache, 2001, Towards a cognitive neuroscience of consciousness: basic evidence and a workspace framework, *Cognition*, 79:1-37.

del Hoyo M., and E. Minian, 2008, Classical invariants for global actions and groupoid atlases, *Applied Categorical Structures*, 18:689-721.

Dembo, A., and O. Zeitouni, 1998, *Large Deviations: Techniques and Applications*, 2nd edition, Springer, New York.

Dietrich, K., R. Douglas, P. Succop, O. Berger, R. Bornschein, 2001, Early exposure to lead and juvenile delinquency, *Neurotoxicology and Teratology*, 23:511-518.

Dohrenwend B.P., and B.S. Dohrenwend, 1974, Social and cultural influences on psychopathology, *Annual Review of Psychology*, 25:417-452.

Dretske, F., 1994, The explanatory role of information, *Philosophical Transactions of the Royal Society, A*, 349:59-70.

Durham, W., 1991, *Coevolution: Genes, Culture and Human Diversity*, Stanford University Press, Palo Alto, CA.

DSMIV, 1994, *Diagnostic and Statistical Manual*, fourth edition, American Psychiatric Association, Washington, DC.

Eaton, W., 1978, Life events, social supports, and psychiatric symptoms: a re-analysis of the New Haven data, *Journal of Health and Social Behavior*, 19:230-234.

Ellis, R., 1985, *Entropy, Large Deviations, and Statistical Mechanics*, Springer, New York.

Emery, M., 1989, *Stochastic Calculus on Manifolds*, Springer, New York.

Feynman, R., 2000, *Lectures on Computation*, Westview Press, New York.

Foley, D., J. Craid, R. Morley, C. Olsson, T. Dwyer, K. Smith, and R. Saffery, 2009, Prospects for epigenetic epidemiology, *American Journal of Epidemiology*, 169:389-400.

Frankel, T., 2006, The Geometry of Physics: An Introduction, Second Edition, Cambridge University Press.

Fullilove, M., 2004, *Root Shock: How Tearing Up City Neighborhoods Hurts America and What we can do about it*, Balantine Books, New York.

Gilbert, P., 2001, Evolutionary approaches to psychopathology: the role of natural defenses, *Australian and New Zealand Journal of Psychiatry*, 35:17-27.

Gilbert, S., 2001, Mechanisms for the environmental regulation of gene expression: ecological aspects of animal development, *Journal of Bioscience*, 30:65-74.

Glass, T., K. Bandeen-Roche, M. McAtee, K. Bolla, A. Todd, B. Schwartz, 2009, Neighborhood psychosocial hazards and the association of cumulative lead dose with cognitive function in older adults, *American Journal of Epidemiology*, 169:683-692.

Glazebrook, J.F., and R. Wallace, 2009a, Small worlds and red queens in the global workspace: an information-theoretic approach, *Cognitive Systems Reserch*, 10:333-365.

Glazebrook, J.F., and R. Wallace, 2009b, Rate distortion manifolds as model spaces for cognitive information, *Informatica*,33:309-345.

Golubitsky, M., and I. Stewart, 2006, Nonlinear dynamics and networks: the groupoid formalism, *Bulletin of the American Mathematical Society*, 43:305-364.

Gunderson, L., 2000, Ecological resilience – in theory and application, *Annual Reviews of Ecological Systematics*, 31:425-439.

Heine, S., 2001, Self as cultural product: an examination of East Asian and North American selves, *Journal of Personality*, 69:881-906.

Helgason, S., 1962, *Differential Geometry and Symmetric Spaces*, Academic Press, New York.

Henrich J., S. Heine, and A. Norenzayan, 2010, The Weirdest people in the world? In press, *Behavioral and Brain Sciences*.

Holling, C., 1973, Resilience and stability of ecological systems, *Annual Reviews of Ecological Systematics*, 4:1-23.

Jablonka, E., and M. Lamb, 1995, *Epigenetic Inheritance and Evolution: The Lamarckian Dimension*, Oxford University Press, Oxford, UK.

Jablonka, E., and M. Lamb, 1998, Epigenetic inheritance in evolution, *Journal of Evolutionary Biology*, 11:159-183.

Jablonka, E., 2004, Epigenetic epidemiology, *International Journal of Epidemiology*, 33:929-935.

Jaenisch, R., and A. Bird, 2003, Epigenetic regulation of gene expression: how the genome integrates intrinsic and environmental signals, *Nature Genetics Supplement*, 33:245-254.

Jacobson J., and Jacobson S. 2002, Breast-feeding and gender as moderators of teratogenic effects on cognitive development, *Neurotoxicological Teratology*, 24:349-358.

Jenkins, J., A. Kleinman, and B. Good, 1990, Cross-cultural studies of depression. In J. Becker and A. Kleinman, *Advances in mood disorders: Theory and Research*, (pp. 67-99), L. Erlbaum, Los Angeles, CA.

Johnson-Laird, P., F. Mancini, and A. Gangemi, 2006, A hyperemotion theory of psychological illnesses, *Psychological Reviews*, 113:822-841.

Karp, R., C. Chen, A. Meyers, 2005, The appearance of discretionary income: influence on the prevalence of under- and over-nutrition, *International Journal of Equity in Health*, 4:10.

Khinchin, A., 1957, *Mathematical Foundations of Information Theory*, Dover, New York.

Kim, J., R. Stewart, S. Kim, et al., 2007, Interactions between life stressors and susceptibility genes (5-HTTLPR and BDNF) on depression in Korean elders, *Biological Psychiatry*, 62:423-428.

Kim, H., D. Sherman, S. Taylor, J. Sasaki, C. Ryu, and J. Xu, 2010, Culture, serotonin receptor polymorphism (5-HTR1A) and locus of attention. In press, *Social, Cognitive, and Affective Neurosciences*.

Kleinman, A., and B. Good, 1985, *Culture and Depression: Studies in the Anthropology of Cross-Cultural Psychiatry of Affect and Depression*, University of California Press, Berkeley, CA.

Kleinman, A., and A. Cohen, 1997, Psychiatry's global challenge, *Scientific American*, 276(3):86-89.

Landau, L., and E. Lifshitz, 2007, *Statistical Physics, 3rd Edition*, Part I, Elsevier, New York.

Manolio, T., F. Collins, N. Cox, et al., 2009, Finding the missing heritability of complex diseases, *Nature*, 461:747-753.

Manson, S., 1995, Culture and major depression: Current challenges in the diagnosis of mood disorders, *Psychiatric Clinics of North America*, 18:487-501.

Markus H., and S. Kitayama, 1991, Culture and the self- implications for cognition, emotion, and motivation, *Psychological Review*, 98:224-253.

Matsuda T., and R. Nisbett, 2006, Culture and change blindness, *Cognitive Science: A Multidisciplinary Journal*, 30:381-399.

Maturana, H., and F. Varela, 1980, *Autopoiesis and Cognition*, Reidel Publishing Company, Dordrecht, Holland.

Maturana, H., and F. Varela, 1992, *The Tree of Knowledge*, Shambhala Publications, Boston, MA.

Marsella, A., 2003, Cultural aspects of depressive experience and disorders. In W. Lonner, D. Dinnel, S, Hays, and D. Sattler (Eds.), *Online Readings in Psychology and Culture* (Unit 9, Chapter 4) (http://www.wwu.edu/ culture), Center for Cross-Cultural Research, Western Washington University, Bellingham, WA.

Miranda, M., D. Kim, M. Overstreet Galeano, C. Paul, A. Hull, S. Morgan. 2007. The relationship between early childhood blood lead levels and performance on end-of-grade tests, *Environmental Health Perspectives*, 115:1242-1247.

Needleman, H., Riess J., Tobin M., Biesecker G., and Greenhouse J., 1996, Bone lead levels and delinquent behavior. *Journal of the American Medical Association*, 275:363-369.

Nesse, R., 2000, Is depression an adaptation?, *Archives of General Psychiatry*, 57:14-20.

Nisbett R., K. Peng, C. Incheol, and A. Norenzayan, 2001, Culture and systems of thought: holistic vs. analytic cognition, *Psychological Review*, 108:291-310.

Norenzayan, A., and S. Heine, 2005, Psychological universals: what are they and how can we know? *Psychological Bulletin*, 131:763-784.

O'Nuallain, S., 2008, Code and context in gene expression, cognition, and consciousness. Chapter 15 in Barbiere, M., (ed.), *The Codes of Life: The Rules of Macroevolution*, Springer, New York, pp. 347-356.

Pettini, M., 2007, *Geometry and Topology in Hamiltonian Dynamics and Statistical Mechanics*, Springer, New York.

Prandolini, R., and M. Moody, 1995, Brownian nature of the Time-Base Error in tape recordings, *Journal of the Audio Engineering Society*, 43:241-247.

Protter, P., 1995, *Stochastic Integration and Differential Equations: A New Approach*, Springer, New York.

Richerson, P., and R. Boyd, 2004, *Not by Genes Alone: How Culture Transformed Human Evolution*, Chicago University Press, Chicago, IL.

Risch, N., R. Herrell, T. Lehner, K. Liang, L. Eaves, J. Hoh, A. Griem, M. Kovacs, J. Ott, and K.R. Merikangas, 2009, Interaction between the serotonin transporter gene (5-HTTLPR), stressful life events, and risk of depression, *Journal of the American Medical Association*, 301:2462-2472.

Rockafellar, R., 1970, *Complex Analysis*, Princeton University Press, Princeton, NJ.

Sarshar, N., and X. Wu, 2007, On Rate-Distortion models for natural images and wavelet coding performance, *IEEE Transactions on Image Processing*, 16:1383-1394.

Sarlio-Lahteenkorva, S., E. Lahelma, 2001, Food insecurity is associated with past and present economic disadvantage and body mass index, *Journal of Nutrition*, 131:2880-2884.

Scherrer, K., and J. Jost, 2007a, The gene and the genon concept: a functional and information-theoretic analysis, *Molecular Systems Biology* 3:87-93.

Scherrer, K., and J. Jost, 2007b, Gene and genon concept: coding versus regulation, *Theory in Bioscience* 126:65-113.

Shankardass, K., McConnell, R., Jerrett, M., Milam, J., Richardson, J., and Berhane, K., 2009, Parental stress increases the effect of traffic-related air pollution on childhood asthma incidence, *Proceedings of the National Academy of Sciences*, 106:12406-12411.

Siliani, G., U. Frith, J. Demonet, F. Fazio, D. Perani, C. Price, C. Frith, and E. Paulesu, 2005, Brain abnormalities underlying altered activation in dyslexia: a vowel based morphometry study, *Brain*, 128(Pt.10):2453-2461.

Skierski, M., A. Grundland, and J. Tuszynski, 1989, Analysis of the three-dimensional time-dependent Landau-Ginzburg equation and its solutions, *Journal of Physics A* (Math. Gen.), 22:3789-3808.

Toulouse, G., S. Dehaene, and J. Changeux, 1986, Spin glass model of learning by selection, *Proceedings of the National Academy of Sciences*, 83:1695-1698.

Turner, B., 2000, Histone acetylation and an epigeneticv code, *Bioessays*, 22:836-845.

Villalobos, M., A. Mizuno, B. Dahl, N. Kemmotsu, and R. Muller, 2005, Reduced functional connectivity between VI and Inferior frontal cortex associated with visimoto performance in autism, *Neuroimage*, 25:916-925.

Wallace, D. and R. Wallace, 1998, Scales of geography, time, and population: the study of violence as a public health problem, *American Journal of Public Health*, 88:1853-1858.

Wallace, D., and R. Wallace, 2000, Life and death in Upper Manhattan and the Bronx: Toward evolutionary perspectives on catastrophic social change, *Environment and Planning A*, 32:1245-1266.

Wallace, D., R. Wallace, V. Rauh, 2003, Community stress, demoralization, and body mass index: evidence for social signal transduction, *Social Science and Medicine*, 56:2467-2478.

Wallace, R., 2005, *Consciousness: A Mathematical Treatment of the Global Neuronal Workspace Model*, Springer, New York.

Wallace, R., 2007, Culture and inattentional blindness, *Journal of Theoretical Biology*, 245:378-390.

Wallace, R., 2008, Developmental disorders as pathological resilience domains, *Ecology and Society*, 13:29 (online).

Wallace, R., and M. Fullilove, 2008, *Collective Consciousness and its Discontents: Institutional Distributed Cognition, Racial Policy, and Public Health in the United States*, Springer, New York.

Wallace, R. and D. Wallace, 2005, Structured psychosocial stress and the US obesity epidemic, *Journal of Biological Systems*, 13:363-384.

Wallace, R., and D. Wallace, 2008, Punctuated equilibrium in statistical models of generalized coevolutionary resilience: how sudden ecosystem transitions can entrain both phenotype expression and Darwinian selection, *Transactions on Computational Systems Biology IX*, LNBI 5121:23-85.

Wallace, R., and D. Wallace, 2009, Code, context, and epigenetic catalysis in gene expression, *Transactions on Computational Systems Biology XI*, LNBI 5750:283-334.

Wallace R.G., and R. Wallace, 2009, Evolutionary radiation and the spectrum of consciousness, *Consciousness and Cognition*, 18:160-167.

Wang, M., and G. Uhlenbeck, 1945, On the theory of the Brownian Motion II, *Reviews of Modern Physics*, 17:323-342.

Weinstein, A., 1996, Groupoids: unifying internal and external symmetry, *Notices of the American Mathematical Association*, 43:744-752.

West-Eberhard, M., 2003, *Developmental Placisticity and Evolution*, Oxford University Press, New York.

West-Eberhard, M., 2005, Developmental plasticity and the origin of species differences, *Proceedings of the National Academy of Sciences*, 102:6543-6549.

Zhu, R., Rebirio, A., Salahub, D., Kaufmann, S., 2007, Studying genetic regulatory networks at the molecular level: delayed reaction stochastic models, *Journal of Theoretical Biology*, 246:725-745.

Chapter 3

Protein folding disorders

3.1 Introduction

The existence of 'global' protein folding and aggregation diseases, in conjunction with the elaborate cellular folding regulatory apparatus associated with the endoplasmic reticulum and other structures (e.g., Scheuner and Kaufman, 2008; Dobson, 2003), makes clear that simple physical 'folding funnel' free energy mechanisms are not fully adequate to describe the process, to understate the matter. This suggests that a more biologically-based model is needed, analogous to Atlan and Cohen's (1998) cognitive paradigm for the immune system. That is, the intractable set of disorders related to protein aggregation and misfolding belies simple mechanistic approaches, although free energy landscape pictures surely capture part of the process. The diseases range from prion illnesses like Creutzfeld-Jakob disease, to amyloid-related dysfunctions like Alzheimer's, Huntington's and Parkinson's diseases, and type 2 diabetes. Misfolding disorders include emphysema and cystic fibrosis. A deeper understanding of protein folding mechanisms, in particular of epigenetic, social, and environmental influences, might contribute to prevention and treatment of these debilitating conditions.

More particularly, the role of epigenetic and environmental factors in type 2 diabetes has long been known (e.g., Zhang et al., 2009; Wallach and Rey, 2009). Haataja et al. (2008), for example, conclude that the islet in type 2 diabetes shows much in common with neuropathology in neurodegenerative diseases where interest is now focused on protein misfolding and aggregation and the diseases are now often referred to as unfolded protein diseases.

Scheuner and Kaufman (2008) likewise examine the unfolded protein response in β cell failure and diabetes. Indeed, their opening paragraph raises the fundamental questions regarding the adequacy of simple energy landscape models of protein folding:

In eukaryotic cells, protein synthesis and secretion are precisely coupled with the capacity of the endoplasmic reticulum (ER) to fold, process, and traffic proteins to the cell surface. These processes are coupled through several signal transduction pathways collectively known as the unfolded protein response [that] functions to reduce the amount of nascent protein that enters the ER lumen, to increase the ER capacity to fold protein through transcriptional upregulation of ER chaperones and folding catalysts, and to induce degradation of misfolded and aggregated protein.

Qiu et al. (2009) address Alzheimer's Disease in much the same fashion:

Alzheimer's dementia is a multifactorial disease in which older age is the strongest risk factor... [that] may partially reflect the cumulative effects of different risk and protective factors over the lifespan, including the complex interactions of genetic susceptibility, psychosocial factors, biological factors, and environmental exposures experienced over the lifespan.

Qiu et al. (2009) explain that mutation effects account for only a small fraction of observed cases, and that the APOE ϵ4 allele – the only established genetic factor for both early and late onset disease – is a *susceptibility* gene, neither necessary nor sufficient for disease onset. They further describe how many of the same factors implicated in diabetes and cardiovascular disease predict onset of Alzheimer's as well: tobacco use, high blood pressure, high serum cholesterol, chronic inflammation, as indexed by a higher level of serum C-reactive protein, and diabetes itself. Highly significant protective factors include high educational and socioeconomic status, regular physical exercise, mentally demanding activities, and significant social engagement.

Similarly, Fillit et al. (2008) find that lifestyle risk factors for cardiovascular disease, such as obesity, lack of exercise, smoking, and certain psychosocial factors, have been associated with an increased risk for cognitive decline and dementia, concluding that current evidence indicates an association between hypertension, dyslipidemia and diabetes and cognitive decline and dementia.

Goldschmidt et al. (2010) describe pathological protein fibrillation as follows:

We found that [protein segments with high fibrillation propensity] tend to be buried or twisted into unfavorable conformations for forming beta sheets... For some proteins a delicate balance between protein folding and misfolding

exists that can be tipped by changes in environment, destabilizing mutations, or even protein concentration...

In addition to the self-chaperoning effects described above,
proteins are also protected from fibrillation during the process of folding by molecular chaperones...

Our genome-wide analysis revealed that self-complementary
segments are found in almost all proteins, yet not all proteins are amyloids. The implication is that chaperoning
effects have evolved to constrain self-complementary segments from interaction with each other.

These processes and mechanisms seem no less examples of chemical cognition than the immune/inflammatory responses that Atlan and
Cohen (1998) describe in terms of an explicit cognitive paradigm, or
that characterizes well-studied neural processes. Our own work (Wallace and Wallace, 2008, 2009) introduces a similar, and highly formal, cognitive paradigm for gene expression whose machinery permits
the natural incorporation of epigenetic and environmental signals via
catalytic mechanisms similar to those of Section 5.3 below. The implication is that progress in understanding, preventing, and treating
protein folding and aggregation disorders now requires introduction of
a biologically-based cognitive paradigm for the folding process itself.

The symmetries and dynamics of protein folding are striking and,
in a local sense, fairly well understood (Dill et al. 2007; Wolynes, 1996;
Onuchic and Wolynes, 2004). Goodsell and Olson (2000) show several
typical examples. More general, but less overtly 'symmetric', conformations, however, involve finite tilings of helices, sheets, and attachment
loops that would seem better described using groupoid methods, following the arguments of Weinstein (1996): As Wolynes (1996) put the
matter, "It is the inexact symmetries of biological molecules that are
most striking".

Anfinsen's (1973) thermodynamic hypothesis has strongly dominated thinking on the subject: the native state of a protein has the
lowest Gibbs free energy, determined by the interaction of the amino
acid sequence and the embedding environment (Wolynes, 1996), with
hydrophobic amino acids driven into the center of the 'native' folded
protein structure. More recent work (e.g., summarized in Lei and
Huang, 2010) suggests that large, complex proteins may have native
configurations representing kinetically accessible, rather than thermodynamically minimal, states. Andre et al. (2008) explore the central insight that "...selection is only likely to operate on primordial complexes
with sufficient initial interaction energy to at least partially overcome
the entropic costs of association of the monomers; evolution can only
optimize a complex that is populated sufficiently to confer a benefit on
the organism".

Here we will attempt to finesse this general perspective by invoking a rate distortion argument applied to the transmitted signal represented by the translation of the genome into the final, evolutionarily driven, condensation of the molten globule of the resulting amino acid string. The argument, an adaptation of Tlusty's (2007) insights regarding the role of rate distortion constraints in evolutionary process, seems fairly direct. It is based on standard material from statistical physics and information theory, using, respectively, average distortion and the rate distortion function itself, as temperature analogs to produce mirror image 'energy' and 'development' pictures of protein folding.

The final step is to mathematically 'weaken', i.e., generalize, the development perspective, using information sources formally dual to the several chemical cognitive processes involved in protein folding. These then, in a sense, engage in a local, multifactorial, coevolutionary interaction whose quasi-stable dynamic states generate products that are, respectively, correct, repaired, eliminated, or misfolded/aggregated proteins. This set of processes is analogous to quasi-stable ecosystem resilience modes, in the sense of Holling (1973) or Gunderson (2000), and apparently subject to punctuated transitions between them consequent on epigenetic or environmental perturbations.

The argument generates a new class of statistical models based on the asymptotic limit theorems of information, in the same sense that regression and other parametric models are based on the Central Limit Theorem, and these should prove useful in data analysis as well as providing a new conceptual approach.

We begin with a restatement of some standard material from statistical physics that provides the basis for a subsequent argument-by-abduction.

3.2   Spontaneous symmetry breaking

Landau's theory of phase transitions (Landau and Lifshitz, 2007), described briefly in the previous chapter, assumes that the free energy of a system near criticality can be expanded in a power series of some 'order parameter' ϕ representing a fundamental measurable quantity, that is, a symmetry invariant. One writes

$$F_0 = \sum_{k=m}^{p(>m)} A_k \phi^k,$$

(3.1)

with $A_2 \approx \alpha(T - T_c)$ sufficiently close to the critical temperature T_c. This mean field approach can be used to describe a variety of second-order effects for $p = 4$ or $p = 6$, $A_3 = 0$ and $A_4 > 0$, and first order phase transitions (requiring latent heat) for either $p = 6, A_3 = 0, A_4 < 0$ or $p = 4$ and $A_3 \neq 0$. These can be both temperature induced (for $m = 2$) and field induced (for $m = 1$).

Minimization of F_0 with respect to the order parameter yields the average value of ϕ, $< \phi >$, which is zero above the critical temperature and non-zero below it. In the absence of external fields, the second-order transition occurs at $T = T_c$, while the first-order, needing latent heat, occurs at $T_c^* = T_c + A_4^2/4\alpha A_6$. In the latter case thermal hysteresis arises between $T_s \equiv T_c + A_4^2/3\alpha A_6$ and T_c. A more accurate approximation involves an expression that recognizes the effect of coarse-graining, adding a term in $\nabla^2\phi$ and integrating over space rather than summing. Regimes dominated by this gradient will show behaviors analogous to those described using the one dimensional Landau-Ginzburg equation, which, among other things, characterizes superconductivity.

The Landau formalism quickly enters deep topological waters (Pettini, 2007, pp. 42-43; Landau and Lifshitz, 2007, pp. 459-466). The essence of Landau's insight was that phase transitions without latent heat – second order transitions – were usually in the context of a significant symmetry change in the physical states of a system, with one phase, at higher temperature, being far more symmetric than the other. A symmetry is lost in the transition, a phenomenon called spontaneous symmetry breaking. The greatest possible set of symmetries in a physical system is that of the Hamiltonian describing its energy states. Usually states accessible at lower temperatures will lack symmetries available at higher temperatures, so that the lower temperature phase is the less symmetric: The randomization of higher temperatures ensures that higher symmetry/energy states will then be accessible to the system.

At the lower temperature an order parameter must be introduced to describe the system's physical states – some extensive quantity like magnetization. The order parameter will vanish at higher temperatures, involving more symmetric states, and will be different from zero in the less symmetric lower temperature phase.

This can be formalized, following Pettini (2007), as follows. Consider a thermodynamic system having a free energy F which is a function of temperature T, pressure P, and some other extensive macroscopic parameters m_i, so that $F = F(P, T, m_i)$. The m_i all vanish in the most symmetric phase, so that, as a function of the m_i, $F(P, T, m_i)$ is invariant with respect to the transformations of the symmetry group G_0 of the most symmetric phase of the system when all $m_i \equiv 0$.

The state of the system can be represented by a vector $|m>=|m_1, ..., m_n>$ in a vector space $\mathcal{E}$. Now, within $\mathcal{E}$, construct a linear representation of the group G_0 that associates with any $g \in G_0$ a matrix $M(g)$ having rank n. In general, the representation $M(g)$ is reducible, and we can decompose $\mathcal{E}$ into invariant irreducible subspaces $\mathcal{E}_1, \mathcal{E}_2, ..., \mathcal{E}_k$, having basis vectors $|e_i^{(n)}>$ with $n = 1, 2, ...n_i$ and $n_j = dim\mathcal{E}_i$. The state variables m_i are transformed into new variables $\eta_i^{(n)} = <e_i^{(n)}|m>$, where the bracket represents an inner product.

In terms of irreducible representations $D_i(g)$ induced by $M(g)$ in $\mathcal{E}_i$ we have

$$M(g) = D_1(g) \oplus D_2(g) \oplus, ..., \oplus D_k(g).$$

If at least one of the $\eta_i^{(n)}$ is nonzero, then the system no longer has the symmetry G_0. This symmetry has been broken, and the new symmetry group is G_i, associated with the representation $D_i(g)$ in $\mathcal{E}_i$. The variables $\eta_i^{(n)}$ are the new order parameters, and the free energy is now $F = F(P, T, \eta_i^{(n)})$. For a physical system the actual values of the η as functions of P and T can be variationally determined by minimizing the free energy F.

Two essential features distinguish information systems, like the translation of a genome into a folded protein, from this simple physical model.

First, the dynamics of order parameters cannot always be determined by simplistic minimization procedures in biological circumstances (e.g., Levinthal, 1969): embedding environments can, within contextual constraints (that particularly include available metabolic free energy), write images of themselves via evolutionary selection mechanisms, driving the system toward such structures as the protein folding funnel (e.g., Levinthal, 1968; Wolynes, 1996).

Second, the essential symmetry of information sources is quite often driven by groupoid, rather than group, structures (e.g., Wallace, 2010). One must then engage the full transitive orbit/isotropy group decomposition, and examine groupoid representations (Bos, 2007; Buneci, 2003) configured about the irreducible representations of the isotropy groups. This observation seems particularly relevant given the usual helix/sheet/connecting loop tilings that characterize most elaborate protein conformations (Wolynes, 1996).

A brief summary of standard material on groupoids is included as a Mathematical Appendix.

3.3   Information theory

Here we think of the machinery listing a sequence of codons as communicating with machinery that produces amino acids, folds them *in context*, and produces the final symmetric protein. We then suppose it possible to compare what is actually produced with what should have been produced, perhaps by a simple evolutionary survival mechanism, perhaps via some more sophisticated error-correcting systems. This is not a new idea, and Onuchic and Wolynes (2004), for example, put the matter fully in evolutionary terms:

> Protein folding should be complex... a folding mechanism must involve a complex network of elementary interactions. However, simple empirical patterns of protein folding kinetics... have been shown to exist.
>
> This simplicity is owed to the global organization of the landscape of the energies of protein conformations into a funnel...This organization is not characteristic of all polymers with any sequence of amino acids, but is a result of evolution...
>
> Evolution achieves robustness by selecting for sequences in which the interactions present in the functionally useful structure are not in conflict, as in a random heteropolymer, but instead are mutually supportive and cooperatively lead to a low energy structure. The interactions are 'minimally frustrated'... or 'consistent'...

It is possible to reframe something of this mechanism in formal information theory terms.

Suppose a sequence of signals is generated by a biological information source Y having output $y^n = y_1, y_2, \ldots$ – codons. This is 'digitized' in terms of the observed behavior of the system with which it communicates, say a sequence of 'observed behaviors' $b^n = b_1, b_2, \ldots$ – amino acids and their folded protein structure. Assume each b^n is then deterministically retranslated back into a reproduction of the original biological signal, $b^n \to \hat{y}^n = \hat{y}_1, \hat{y}_2, \ldots$.

Define a distortion measure $d(y, \hat{y})$ which compares the original to the retranslated path. Many distortion measures are possible. The Hamming distortion is defined simply as

$$d(y, \hat{y}) = 1, y \neq \hat{y}$$

$$d(y, \hat{y}) = 0, y = \hat{y}.$$

For continuous variates the squared error distortion is just $d(y, \hat{y}) = (y - \hat{y})^2$.

There are many such possibilities. The distortion between *paths* y^n and $\hat{y}^n$ is defined as $d(y^n, \hat{y}^n) \equiv \frac{1}{n} \sum_{j=1}^{n} d(y_j, \hat{y}_j)$.

A remarkable fact of the Rate Distortion Theorem is that *the basic result is independent of the exact distortion measure chosen* (Cover and Thomas, 1991; Dembo and Zeitouni, 1998).

Suppose that with each path y^n and b^n-path retranslation into the y-language, denoted $\hat{y}^n$, there are associated individual, joint, and conditional probability distributions $p(y^n), p(\hat{y}^n), p(y^n, \hat{y}^n), p(y^n|\hat{y}^n)$.

The average distortion is defined as

$$D \equiv \sum_{y^n} p(y^n)d(y^n, \hat{y}^n).$$

(3.2)

It is possible, using the distributions given above, to define the information transmitted from the Y to the $\hat{Y}$ process using the Shannon source uncertainty of the strings:

$$I(Y, \hat{Y}) \equiv H(Y) - H(Y|\hat{Y}) = H(Y) + H(\hat{Y}) - H(Y, \hat{Y}),$$

where $H(..., ...)$ is the standard joint, and $H(...|...)$ the conditional, Shannon uncertainties (Cover and Thomas, 1991; Ash, 1990).

If there is no uncertainty in Y given the retranslation $\hat{Y}$, then no information is lost, and the systems are in perfect synchrony.

In general, of course, this will not be true.

The *rate distortion function* $R(D)$ for a source Y with a distortion measure $d(y, \hat{y})$ is defined as

$$R(D) = \min_{p(y,\hat{y}); \sum_{(y,\hat{y})} p(y)p(y|\hat{y})d(y,\hat{y}) \leq D} I(Y, \hat{Y}).$$

(3.3)

The minimization is over all conditional distributions $p(y|\hat{y})$ for which the joint distribution $p(y, \hat{y}) = p(y)p(y|\hat{y})$ satisfies the average distortion constraint (i.e., average distortion $\leq D$).

The *Rate Distortion Theorem* states that $R(D)$ is the minimum necessary rate of information transmission which ensures the communication between the biological vesicles does not exceed average distortion D. Thus $R(D)$ defines a minimum necessary channel capacity. Cover and Thomas (1991) or Dembo and Zeitouni (1998) provide details. The rate distortion function has been calculated for a number of systems.

We reiterate an absolutely central fact characterizing the rate distortion function: Cover and Thomas (1991, Lemma 13.4.1) show that *R(D) is necessarily a decreasing convex function of D for any reasonable definition of distortion.*

That is, *R(D) is always* a reverse J-shaped curve. This will prove crucial for the overall argument. Indeed, convexity is an exceedingly powerful mathematical condition, and permits deep inference (e.g., Rockafellar, 1970). Ellis (1985, Ch. VI) applies convexity theory to conventional statistical mechanics.

For a Gaussian channel having noise with zero mean and variance σ^2 (Cover and Thomas, 1991),

$$R(D) = 1/2 \log[\sigma^2/D], 0 \leq D \leq \sigma^2$$

$$R(D) = 0, D > \sigma^2.$$

(3.4)

Recall, now, the relation between information source uncertainty and channel capacity (e.g., Ash, 1990):

$$H[X] \leq C,$$

(3.5)

where H is the uncertainty of the source X and C the channel capacity, defined according to the relation (Ash, 1990)

$$C \equiv \max_{P(X)} I(X|Y),$$

(3.6)

where $P(X)$ is chosen so as to maximize the rate of information transmission along a channel Y.

Note that for a parallel set of noninteracting channels, the overall channel capacity is the sum of the individual capacities, providing a powerful 'consensus average' that does not apply in the case of modern molecular coding.

Finally, recall the analogous definition of the rate distortion function above, again an extremum over a probability distribution.

Our own work (Wallace and Wallace, 2008) focuses on the homology between information source uncertainty and free energy density. More formally, if $N(n)$ is the number of high probability 'meaningful' – that is, grammatical and syntactical – sequences of length n emitted by an information source X, then, according to the Shannon-McMillan Theorem, the zero-error limit of the Rate Distortion Theorem (Ash, 1990; Cover and Thomas, 1991; Khinchin, 1957),

$$H[X] = \lim_{n \to \infty} \frac{\log[N(n)]}{n}$$

$$= \lim_{n \to \infty} H(X_n | X_0, ..., X_{n-1})$$

$$= \lim_{n \to \infty} \frac{H(X_0, ..., X_n)}{n+1},$$

(3.7)

where, again, $H(...|...)$ is the conditional and $H(..., ...)$ is the joint Shannon uncertainty.

In the limit of large n, $H[X]$ becomes homologous to the free energy density of a physical system at the thermodynamic limit of infinite volume. More explicitly, the free energy density of a physical system having volume V and partition function $Z(\beta)$ derived from the system's Hamiltonian – the energy function – at inverse temperature β is (e.g., Landau and Lifshitz 2007)

$$F[K] = \lim_{V \to \infty} -\frac{1}{\beta} \frac{\log[Z(\beta, V)]}{V} \equiv$$

$$\lim_{V \to \infty} \frac{\log[\hat{Z}(\beta, V)]}{V},$$

with $\hat{Z} = Z^{-1/\beta}$. The latter expression is formally similar to the first part of equation (7), a circumstance having deep implications: Feynman (2000) describes in great detail how information and free energy have an inherent duality. Feynman, in fact, defines information precisely as the free energy needed to erase a message. The argument is surprisingly direct (e.g., Bennett, 1988), and for very simple systems it is easy to design a small (idealized) machine that turns the information within a message directly into usable work – free energy. Information is a form of free energy and the construction and transmission of information within living things consumes metabolic free energy, with nearly inevitable losses via the second law of thermodynamics. If there are limits on available metabolic free energy there will necessarily be limits on the ability of living things to process information.

Figure 3.1 presents a schematic of the mechanism: As the complexity of a dynamic physiological information process rises, that is, as H increases, its free energy content increases linearly. The metabolic free energy needed to construct and maintain the physiological systems that instantiate H should, however, be expected to increase nonlinearly with it, hence the 'translation gap' of the figure. Section 5 of Wallace (2010) gives a fairly elementary derivation of such a relation in terms of rate distortion theory. Figure 3.1 suggests that H may indeed be a good, if highly nonlinear, index of large-scale free energy dynamics.

Conversely, information source uncertainty has an important heuristic interpretation that Ash (1990) describes as follows:

> [W]e may regard a portion of text in a particular language as being produced by an information source. The probabilities $P[X_n = a_n | X_0 = a_0, ... X_{n-1} = a_{n-1}]$ may be estimated from the available data about the language; in this way we can estimate the uncertainty associated with the language. A large uncertainty means, by the [Shannon-McMillan Theorem], a large number of 'meaningful' sequences. Thus given two languages with uncertainties H_1 and H_2 respectively, if $H_1 > H_2$, then in the absence of noise it is easier to communicate in the first language; more can be said in the same amount of time. On the other hand, it will be easier to reconstruct a scrambled portion of text in the second language, since fewer of the possible sequences of length n are meaningful.

In sum, if a biological system characterized by H_1 has a richer and more complicated internal communication structure than one characterized by H_2, then necessarily $H_1 > H_2$ and system 1 represents a more energetic process than system 2, and by the arguments of figure 3.1, may trigger even greater metabolic free energy dynamics, as is shown in more detail in Chapter 4.

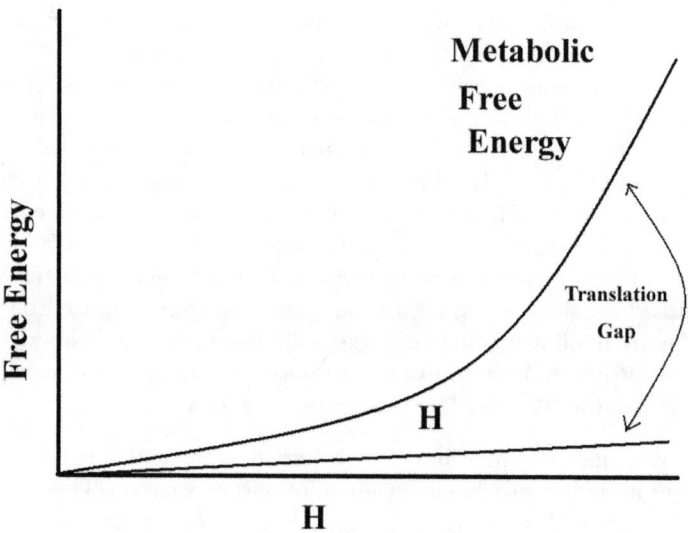

Figure 3.1: Nonlinear increase in metabolic free energy needed to maintain and generate linear increase in the information source uncertainty of a complex physiological process. H is seen to 'leverage' metabolic expenditures, parameterizing a more complicated nonequilibrium thermodynamics. See Chapter 4 for an explicit calculation in a somewhat different system.

By equations (3.5), (3.6), and (3.7), the Rate Distortion Function, $R(D)$ is likewise a free energy measure, constrained by the availability of metabolic free energy.

3.4 The energy picture

Ash's comment leads directly to a model in which the average distortion between the initial codon stream and the final form of the folded amino acid stream, the protein, becomes a dominant force, particularly in an evolutionary context in which fidelity of codon expression has survival value. The direct model examines the distortion between the codon stream and the folded protein structure.

Suppose there are n possible folding schemes. The most familiar approach, perhaps, is to assume that a given distortion measure, D, under evolutionary selection constraints, serves much as an external temperature bath for the possible distribution of conformation free energies, the set $\{\mathcal{H}_1, ..., \mathcal{H}_n\}$. That is, high distortion, represented by a low rate of transmission of information between codon machine and amino acid/protein folding machine, permits a larger distribution of possible symmetries – the big end of the folding funnel – according to the classic formula

$$Pr[\mathcal{H}_j] = \frac{\exp[-\mathcal{H}_j/\lambda D]}{\sum_{i=1}^{n} \exp[-\mathcal{H}_i/\lambda D]},$$

(3.8)

where $Pr[\mathcal{H}_j]$ is the probability of folding scheme j having conformational free energy $\mathcal{H}_j$.

We are, in essence, assuming that $Pr[\mathcal{H}_j]$ is a one parameter distribution in the 'intensive' quantity D.

The free energy Morse Function associated with this probability is

$$F_R = -\lambda D \log[\sum_{i=1}^{n} \exp[-\mathcal{H}_i/\lambda D]].$$

(3.9)

Applying a spontaneous symmetry breaking argument to F_R generates topological transitions in folded protein structure as the 'temperature' D decreases, i.e., as the average distortion declines. That is, as the channel capacity connecting codon machines with amino acid/protein folding machines increases, the system is driven to a particular conformation, according to the 'protein folding funnel'.

3.5   The developmental picture

The developmental approach of Wallace and Wallace (2009) permits a different perspective on protein folding.

We now are concerned with developmental pathways in a 'phenotype space' that, in a series of steps, take the amino acid string $\mathbf{S}_0$ at time 0 to the final folded conformation $\mathbf{S}_f$ at some time t in a long series of distinct, sequential, intermediate configurations $\mathbf{S}_i$.

Let $N(n)$ be the number of possible paths of length n that lead from $\mathbf{S}_0$ to $\mathbf{S}_f$. The essential assumptions are:

[1] This is a highly systematic process governed by a 'grammar' and 'syntax' driven by the folding funnel, so that it is possible to divide all possible paths $x_n = \{\mathbf{S}_0, \mathbf{S}_1, ..., \mathbf{S}_n\}$ into two sets, a small, high probability subset that conforms to the demands of the folding funnel topology, and a much larger 'nonsense' subset having vanishingly small probability.

[2] If $N(n)$ is the number of high probability paths of length n, then the 'ergodic' limit

$$H = \lim_{n \to \infty} \log[N(n)]/n$$

(3.10)

both exists and is independent of the path x. This is, essentially, a restatement of the Shannon-McMillan Theorem (Khinchin, 1957).

That is, the folding of a particular protein, from its amino acid string to its final form, is not a random event, but represents a highly – evolutionarily – structured (i.e., by the folding funnel) 'statement' by an information source having source uncertainty H.

3.5.1   Symmetry arguments

A formal equivalence class algebra can now be constructed by choosing different origin and end points $\mathbf{S}_0, \mathbf{S}_f$ and defining equivalence of two states by the existence of a high probability meaningful path connecting them with the same origin and end. Disjoint partition by equivalence class, analogous to orbit equivalence classes for dynamical systems, defines the vertices of the proposed network of developmental protein 'languages'. We thus envision a *network of metanetworks*. Each vertex then represents a different equivalence class of developmental information sources. This is an abstract set of metanetwork 'languages'.

This structure generates a groupoid, in the sense of Weinstein (1996). States a_j, a_k in a set A are related by the groupoid morphism if and only if there exists a high probability grammatical path connecting them to the same base and end points, and tuning across the various possible ways in which that can happen – the different developmental languages – parameterizes the set of equivalence relations and creates the (very large) groupoid.

There is an implicit hierarchy. First, there is structure *within the system having the same base and end points*. Second, there is a complicated groupoid structure defined by sets of dual information sources surrounding the variation of base and end points. We do not need to know what that structure is in any detail, but can show that its existence has profound implications.

We begin with the simple case, the set of dual information sources associated with a fixed pair of beginning and end states.

The first level

Taking the serial grammar/syntax model above, we find that not all high probability meaningful paths from $\mathbf{S}_0$ to $\mathbf{S}_f$ are actually the same. They are structured by the uncertainty of the associated dual information source, and that has a homological relation with free energy density.

Let us index possible information sources connecting base and end points by some set $A = \cup \alpha$. Argument by abduction from statistical physics is direct. The minimum channel capacity needed to produce average distortion less than D in the energy picture above is $R(D)$. We take the probability of a particular H_α as determined by the standard expression

$$P[H_\beta] = \frac{\exp[-H_\beta/\mu R]}{\sum_\alpha \exp[-H_\alpha/\mu R]},$$

(3.11)

where the sum may, in fact, be a complicated abstract integral.

A basic requirement, then, is that the sum/integral always converges.

Thus, in this formulation, there must be structure *within* a (cross sectional) connected component in the base configuration space, determined by R. Some dual information sources will be 'richer'/smarter than others, but, conversely, must use more available channel capacity for their completion.

The second level

While we might simply impose an equivalence class structure based on equal levels of energy/source uncertainty, producing a groupoid (and possibly allowing a Morse Theory approach), we can do more *by now allowing both source and end points to vary*, as well as by imposing energy-level equivalence. This produces a far more highly structured groupoid.

Equivalence classes define groupoids, by standard mechanisms. The basic equivalence classes – here involving both information source uncertainty level and the variation of S_0 and S_f, will define transitive groupoids, and higher order systems can be constructed by the union of transitive groupoids, having larger alphabets that allow more complicated statements in the sense of Ash above.

Again, given a minimum necessary channel capacity R, we propose that the metabolic-energy-constrained probability of an information source representing equivalence class G_i, H_{G_i}, will again be given by

$$P[H_{G_i}] = \frac{\exp[-H_{G_i}/\kappa R]}{\sum_j \exp[-H_{G_j}/\kappa R]},$$

(3.12)

where the sum/integral is over all possible elements of the largest available symmetry groupoid. By the arguments of Ash above, compound sources, formed by the union of underlying transitive groupoids, being more complex, generally having richer alphabets, as it were, will all have higher free-energy-density-equivalents than those of the base (transitive) groupoids.

Let

$$Z_G = \sum_j \exp[-H_{G_j}/\kappa R].$$

(3.13)

We now define the *Groupoid free energy* of the system, a Morse Function F_G, at channel capacity R, as

$$F_G[R] = -\frac{1}{\kappa R} \log[Z_G[R]].$$

(3.14)

These free energy constructs permit introduction of the spontaneous symmetry breaking arguments above, but now an *increase* in R (with corresponding decrease in average distortion D) permits richer system dynamics – higher source uncertainty – resulting in more rapid transmission of the 'message' constituting convergence from $\mathbf{S}_0$ to $\mathbf{S}_f$.

3.5.2   Folding speed and mechanism

Dill et al. (2007) describe the conundrum of folding speeds as follows:

> ...[P]rotein folding speeds – now known to vary over more than eight orders of magnitude – correlate with the topology of the native protein: fast folders usually have mostly local structure, such as helices and tight turns, whereas slow folders usually have more non-local structure, such as β sheets (Plaxco et al., 1998)...

A simple groupoid probability argument reproduces this result. Assume that protein structure can be characterized by some groupoid representing, at least, the disjoint union of the groups describing the symmetries of component secondary structures – helices and sheets. Then, in equation 3.11, the set $A = \cup \alpha$ grows in size – cardinality – with increasing structural complexity. If channel capacity is capped by some mechanism, so that (at least) R grows at a lesser rate than A, by some measure, then

$$P[H_\beta] = \frac{\exp[-H_\beta/\mu R]}{\sum_\alpha \exp[-H_\alpha/\mu R]}$$

must decrease with increase in the number of possible states α, i.e., with increase in the cardinality of R, producing progressively lower rates of convergence to the final state.

These matters lead to the next central question: How can folding rates be modulated?

3.5.3   Catalysis of protein folding

Incorporating the influence of embedding contexts – epigenetic or chaperone effects, or the effects of (broadly) toxic exposures – can be done here by invoking the Joint Asymptotic Equipartition Theorem (JAEPT)

(Cover and Thomas, 1991). For example, given an embedding contextual information source, say Z, that affects protein development, then the developmental source uncertainty H_{G_i} is replaced by a joint uncertainty $H(X_{G_i}, Z)$. The objects of interest then become the jointly typical dual sequences $y^n = (x^n, z^n)$, where x is associated with protein folding development and z with the embedding context. Restricting consideration of x and z to those sequences that are in fact jointly typical allows use of the information transmitted from Z to X as the splitting criterion.

One important inference is that, from the information theory 'chain rule' (Cover and Thomas, 1991),

$$H(X,Y) = H(X) + H(Y|X) \leq H(X) + H(Y),$$

while there are approximately $\exp[nH(X)]$ typical X sequences, and $\exp[nH(Z)]$ typical Z sequences, and hence $\exp[n(H(x) + H(Y))]$ independent joint sequences, there are only about

$$\exp[nH(X,Z)] \leq \exp[n(H(X) + H(Y))]$$

jointly typical sequences, so that the effect of the embedding context, in this model, is to lower the *relative* free energy of a particular protein channel.

Thus the effect of epigenetic/catalytic regulation or toxic exposure is to channel protein into pathways that might otherwise be inhibited or slowed by an energy barrier. Hence the epigenetic/catalytic/toxic information source Z acts as a *tunable catalyst*, a kind of second order enzyme, to enable and direct developmental pathways. This result permits hierarchical models similar to those of higher order cognitive neural function (Wallace, 2005).

This is indeed a relative energy argument, since, metabolically, two systems must now be supported, i.e., that of the 'reaction' itself and that of its catalytic regulator. 'Programming' and stabilizing inevitably intertwined, as it were.

Protein folding, in the developmental picture, can be visualized as a series of branching pathways. Each branch point is a developmental decision, or switch point, governed by some regulatory apparatus (if only the slope of the folding funnel) that may include the effects of toxins or epigenetic mechanisms.

A more general picture emerges by allowing a distribution of possible 'final' states $\mathbf{S}_f$. Then the groupoid arguments merely expand to permit traverse of both initial states and possible final sets, recognizing that there can now be a possible overlap in the latter, and the catalytic effects are realized through the joint uncertainties $H(X_{G_i}, Z)$, so that the guiding information source Z serves to direct as well the possible final states of X_{G_i}.

3.5.4   Extending the model

The most natural extension of the developmental model of protein folding would be in terms of the directed homotopy classification of ontological trajectories, in the sense of Wallace and Wallace (2008, 2009). That is, developmental trajectories themselves can be classified into equivalence classes, for example those that lead to a normal final state $\mathbf{S}_f$, and those that lead to pathological aggregations or misfoldings, say some set $\{\mathbf{S}^i_{path}\}, i = 1, 2, ...$ This produces a dynamic directed homotopy groupoid topology whose understanding might be useful across a broad spectrum of diseases.

Figure 3.2 illustrates the concept. The initial developmental state $\mathbf{S}_0$ can, in this picture, 'fall' down two different sets of developmental pathways, separated by a critical period 'shadow' preventing crossover between them. Paths within one set can be topologically transformed into each other without crossing the filled triangle, and constitute a directed homotopy equivalence classes. The lower apex of the triangle can, however, start at many possible critical period points along any path connecting $\mathbf{S}_0$ and $\mathbf{S}_f$, following the arguments of Section 12 of Wallace and Wallace (2009).

Onset of a path that converges on the conformation $\mathbf{S}_{path}$ is, according to the model, driven by a genetic, epigenetic, or environmental catalysis event, in the sense of Section 3.5.3. The topological equivalence classes define a groupoid on the developmental system.

3.6   A cognitive paradigm

We now take the developmental perspective as the foundation for generating an empirically-based statistical model – effectively a cognitive paradigm for protein folding – that incorporates the embedding contexts of epigenetic and environmental signals. Atlan and Cohen (1998),

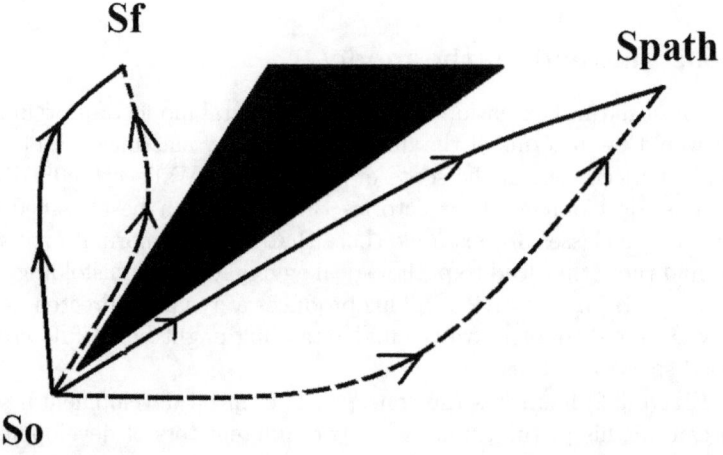

Figure 3.2: Given an initial state S_0 and a critical period casting a path-dependent developmental shadow, there are two different directed homotopy equivalence classes of deformable paths leading, respectively, to the normal folded protein state S_f and the pathological state S_{path}. These sets of paths form equivalence classes defining a topological groupoid.

in the context of a study of the immune system, argue that the essence of cognition is the comparison of a perceived signal with an internal, learned picture of the world, and then choice of a single response from a large repertoire of possible responses. Such choice inherently involves information and information transmission since it always generates a reduction in uncertainty, as explained in Ash (1990, p. 21). Thus structures that process information are constrained by the asymptotic limit theorems of information theory, in the same sense that sums of stochastic variables are constrained by the Central Limit Theorem, allowing the construction of powerful statistical tools useful for data analysis.

More formally, a pattern of incoming input $\mathbf{S}_i$ describing the folding status of the protein – starting with the initial codon stream $\mathbf{S}_0$ of equation Section 5 – is mixed in a systematic algorithmic manner with a pattern of otherwise unspecified 'ongoing activity', including cellular, epigenetic and environmental signals, $\mathbf{W}_i$, to create a path of combined signals $x = (a_0, a_1, ..., a_n, ...)$. Each a_k thus represents some functional composition of internal and external factors, and is expressed in terms of the intermediate states as

$$\mathbf{S}_{i+1} = f([\mathbf{S}_i, \mathbf{W}_i]) = f(a_i)$$

(3.15)

for some unspecified function f. The a_i are seen to be very complicated composite objects, in this treatment that we may choose to coarse-grain so as to obtain an appropriate 'alphabet'.

In a simple spinglass-like model, $\mathbf{S}$ would be a vector, $\mathbf{W}$ a matrix, and f would be a function of their product at 'time' i.

The path x is fed into a highly nonlinear decision oscillator, h, a 'sudden threshold machine' pattern recognition structure, in a sense, that generates an output $h(x)$ that is an element of one of two disjoint sets B_0 and B_1 of possible system responses. Let us define the sets B_k as

$$B_0 = \{b_0, ..., b_k\},$$

$$B_1 = \{b_{k+1}, ..., b_m\}.$$

Assume a graded response, supposing that if $h(x) \in B_0$, the pattern is not recognized, and if $h(x) \in B_1$, the pattern has been recognized, and some action $b_j, k + 1 \le j \le m$ takes place. Typically, the set B_1

would represent the final state of the folded protein, either normal or in some pathological conformation, that is sent on in the biological process or else subjected to some attempted corrective action. Corrections may, for example, range from activation of 'heat shock' protein repair to more drastic clean-up attack.

The principal objects of formal interest are paths x triggering pattern recognition-and-response. That is, given a fixed initial state $a_0 = [\mathbf{S}_0, \mathbf{W}_0]$, examine all possible subsequent paths x beginning with a_0 and leading to the event $h(x) \in B_1$. Thus $h(a_0, ..., a_j) \in B_0$ for all $0 < j < m$, but $h(a_0, ..., a_m) \in B_1$. B_1 is thus the set of final possible states, $\{\mathbf{S}_f\} \cup \{\mathbf{S}_{path}\}$ from figure 3.2 that includes both the final 'physics' state $\mathbf{S}_f$ and the set of possible pathological conformations.

Again, for each positive integer n, let $N(n)$ be the number of high probability grammatical and syntactical paths of length n which begin with some particular a_0 and lead to the condition $h(x) \in B_1$. Call such paths 'meaningful', assuming, not unreasonably, that $N(n)$ will be considerably less than the number of all possible paths of length n leading from a_0 to the condition $h(x) \in B_1$.

While the combining algorithm, the form of the nonlinear oscillator, and the details of grammar and syntax, can all be unspecified in this model, the critical assumption that permits inference of the necessary conditions constrained by the asymptotic limit theorems of information theory is that the finite limit

$$H = \lim_{n \to \infty} \frac{\log[N(n)]}{n}$$

(3.16)

both exists and is independent of the path x.

Call such a pattern recognition-and-response cognitive process *ergodic*. Not all cognitive processes are likely to be ergodic in this sense, implying that H, if it indeed exists at all, is path dependent, although extension to nearly ergodic processes seems possible (Wallace and Fullilove, 2008).

Invoking the spirit of the Shannon-McMillan Theorem, as choice involves an inherent reduction in uncertainty, it is then possible to define an adiabatically, piecewise stationary, ergodic (APSE) information source $\mathbf{X}$ associated with stochastic variates X_j having joint and conditional probabilities $P(a_0, ..., a_n)$ and $P(a_n | a_0, ..., a_{n-1})$ such that appropriate conditional and joint Shannon uncertainties satisfy the classic relations of equation (3.7).

This information source is defined as *dual* to the underlying ergodic cognitive process.

Adiabatic means that the source has been parameterized according to some scheme, and that, over a certain range, along a particular piece, as the parameters vary, the source remains as close to stationary and ergodic as needed for information theory's central theorems to apply. *Stationary* means that the system's probabilities do not change in time, and *ergodic*, roughly, that the cross sectional means approximate long-time averages. Between pieces it is necessary to invoke various kinds of phase transition formalisms, as described more fully in e.g., Wallace (2005).

Structure is now subsumed *within the sequential grammar and syntax of the dual information source* rather than within the set of developmental paths of figure 3.2 and the added catalysis arguments of Section 3.5.3.

This transformation in perspective carries heavy computational burdens, as well as providing deeper mathematical insight, as cellular machineries, and phenomena of epigenetic or environmental catalysis, are now included within a single model.

The energy and development pictures of Sections 3.4 and 3.5 were 'dual' as simply different aspects of the convexity of the rate distortion function with average distortion. This model seems qualitatively different, as we are now invoking a 'black box' information theory statistical model involving grammar and syntax driven by an asymptotic limit theorem, the Shannon-McMillan Theorem. The set of nonequilibrium empirical generalized Onsager models derived from it, as in Wallace and Wallace (2008, 2009), is based on the information source uncertainty H as a free energy-analog (Wallace and Wallace, 2009), thus having a significantly different meaning from those above, and are more similar to regression models fitted according to the Central Limit Theorem. In a manner similar to the treatment in Wallace (2005), the system becomes subject to 'biological' renormalizations at critical, highly punctuated, transitions.

The most evident assumption at this point is that there may be more than a single cognitive protein folding process in operation, e.g., that the action of chaperones and other corrective mechanisms involve separate cognitive processes $\{H_1, ..., H_m\}$ that interact via some form of crosstalk. Following the direction of Wallace and Wallace (2009) we invoke a complicated version of an internal system of empirical Onsager relations, assuming that the different cognitive processes represented by these dual information sources *become each others primary environments*, a broadly, if locally, coevolutionary phenomenon, in the sense of Diekmann and Law (1996). We write

$$H_k = H_k(K_1, ..., K_s, ..., H_j, ...)$$

(3.17)

where the K_s represent other relevant parameters and $k \neq j$. In a generalization of the statistical model, we would expect the dynamics of such a system to be driven by an empirical recursive network of stochastic differential equations. Letting the K_s and H_j all be represented as parameters Q_j, with the caveat that H_k not depend on itself, we are able to define an entropy-analog based on the homology of information source uncertainty with free energy as

$$S_k = H_k - \sum_i Q_i \partial H_k / \partial Q_i,$$

(3.18)

whose gradients in the Q define local (broadly) chemical forces. In close analogy with other nonequilibrium phenomena we obtain a complicated recursive system of phenomenological Onsager relation stochastic differential equations:

$$dQ_t^j = \sum_i [L_{j,i}(t, ..., Q_k, ...)dt + \sigma_{j,i}(t, ..., Q_k, ...)dB_t^i],$$

(3.19)

where, again, for notational simplicity, we have expressed both parameters and information sources in terms of the same symbols Q^k. The dB_t^i represent different kinds of 'noise' having particular forms of quadratic variation that may represent a projection of environmental factors under something like a rate distortion manifold (Glazebrook and Wallace, 2009a, b).

The Mathematical Appendix provides an introduction to stochastic differential equations via the Martingale Theorem.

As usual for such systems, there can be multiple quasi-stable points within a given system's $\{..., H_k..., ..., K_j, ...\}$ representing a class of generalized resilience modes (Holling, 1973; Gunderson, 2000; Wallace and Wallace, 2008) accessible via punctuation as various possible outcomes of the protein folding process: normal, repaired, eliminated, and pathological. These states can, in theory, be found by setting equation 19 to zero, as the noise terms preclude unstable equilibria. As described elsewhere (Diekmann and Law, 1996; Champagnat et al., 2006; Wallace and Wallace 2009), however, far more complicated 'coevolutionary' behaviors can be expected that we will not explore further: here we enter deep biological waters whose exploration will require a significant extension of our general formal perspective. Glazebrook and Wallace (2009a, b) provide something of a mathematical roadmap.

The essential point is that, under resilience theory, 'perturbations' of various sorts can be expected to shift the system between different quasi-stable folding modes, and once shifted, correction may be exceedingly difficult or impossible, as these are, broadly, developmental processes having significant path dependence.

3.7 Aging and protein folding

The developmental perspective above, although focused on the relatively short time frames of protein metabolism – in the range from microseconds to minutes – is suggestive. The principal 'risk factor' for a large array of protein folding disorders is biological age – for humans, in the range of decades – and a simplified version of the previous section may provide a life-course perspective, that is, a developmental model over a far longer timescale.

Equations 3.3-3.7 suggest that the rate distortion function, $R(D)$, is itself a free energy measure, as it represents the minimum channel capacity needed to assure average distortion equal to or less than D. Let us now consider the principal branch in figure 3.2, the set of paths from S_0 to S_f, representing normal protein folding, taken as a communication channel having a given rate distortion function. The arguments of the previous section suggest that there will be an empirical Onsager relation in the gradient of the *rate distortion disorder*, an entropy-analog,

$$S_R \equiv R(D) - DdR(D)/dD$$

(3.20)

such that, over a life-history timeline,

$$dD/dt = f(dS_R/dD)$$

(3.21)

for some appropriate function f.

For a Gaussian channel, having $R(D) = (1/2)\log(\sigma^2/D)$, $S_R(D) = (1/2)\log(\sigma^2/D) + 1/2$, the simplest possible Onsager relation becomes

$$dD/dt = -\mu dS_R dD = \mu/2D,$$

(3.22)

with the explicit solution

$$D = \sqrt{\mu t}.$$

(3.23)

For an appropriate timescale – necessarily many orders of magnitude longer than the time of folding itself – the average distortion, representing the degree of misfolding, simply grows as a diffusion process in time. This is the simplest possible aging model, in which μ represents the accumulated impacts of epigenetic and broadly environmental effects including toxic exposures, nutrition, the richness of social interaction, and so on, over a lifetime.

A somewhat less simplistic model takes the Onsager relation as constrained by the availability of metabolic free energy, M, that powers active chaperone processes,

$$dD/dt = -\mu dS_R/dD - \kappa M = \mu/2D - \kappa M$$

(3.24)

where κ represents the efficiency of use of metabolic energy. This equation has the equilibrium solution (when $dD/dt = 0$)

$$D_{equlib} = \mu/2\kappa M.$$

(3.25)

Here aging is represented by a decay in the efficiency of those chaperone processes, i.e., a slow decline in κ, that may involve idiosyncratic dynamics, ranging from punctuated phase transitions to autocatalytic runaway effects, since D, in equation 3.8, acts as a temperature analog for a system able to undergo symmetry breaking.

3.8   Discussion and conclusions

The fidelity of the translation between genome and final protein conformation, measured by an average distortion measure, or its dual, the minimum channel capacity needed to limit average distortion to a given level, serve as evolutionarily-sculpted temperature analogs, in the sense of Onuchic and Wolynes (2004), to determine the possible phase transitions defining different degrees of protein symmetry. The protein folding funnel follows a spontaneous symmetry breaking mechanism with average distortion as the temperature analog, or, in the developmental picture, greater channel capacity leads more directly to the final state $\mathbf{S}_f$.

The various outcomes to the full protein folding process – normal, corrected, eliminated, pathological – emerge, in the expanded 'Onsager relation' statistical model based on a cognitive paradigm, as distinct 'resilience' modes of a complicated internal cellular ecosystem, subject to punctuated transitions driven, in some cases, by signals from embedding epigenetic and ecological structures. Increase in the rate of folding disorders with age emerges through a long-time generalization of the Onsager model.

In a sense this work extends Tlusty's (2007) elegant topological exploration of the evolution of the genetic code, suggesting that rate distortion considerations are central to a broad spectrum of molecular biological phenomena, although different measures may come to the fore under different perspectives.

The cognitive paradigm introduced here opens a unified biological vision of protein folding and its disorders that may relate the etiology of a large set of misfolding and aggregation diseases more clearly to both cellular and epigenetic processes and environmental stressors. This would be, in the current reductionist sandstorm, no small thing. A cognitive paradigm subsumes epigenetic and environmental catalysis of protein conformation 'development' within a single grammar and syntax, and allows both normal folding and its pathologies to both be viewed as 'natural' outcomes, a perspective more consistent with rates of folding and aggregation disorders observed within an aging population.

Most basically, however, such a cognitive paradigm, as we have constructed it, will likely serve as the foundation for a new class of statistical tools – based on the asymptotic limit theorems of information theory rather than on the Central Limit Theorem alone – that should be useful in the analysis of data related to protein misfolding and aggregation disorders.

3.9 References

Andre, I., C. Strauss, D. Kaplan, P. Bradley, and D. Baker, 2008, Emergence of symmetry in homooligomeric biological assemblies, *Proceedings of the National Academy of Sciences*, 105:16148-16152.

Anfinsen, C., 1973, Principles that govern the folding of protein chains, *Science*, 181:223-230.

Ash, R., 1990, *Information Theory*, Dover, New York.

Atlan, H., and I. Cohen, 1998, Immune information, self-organization, and meaning, *International Immunology*, 10:711-717.

Bennett, C., 1988, Logical depth and physical complexity. In Herkin, R. (ed.), *The Universal Turing Machine: A Half-Century Survey*, Oxford University Press, pp. 227-257.

Bos, R., 2007, Continuous representations of groupoids. arXiv:math/0612639.

Brown, R., 1987, From groups to groupoids: a brief survey, *Bulletin of the London Mathematical Society*, 19:113-134.

Buneci, M., 2003, *Representare de Groupoizi*, Editura Mirton, Timosoara, Romania.

Cannas Da Silva, A., and Weinstein, A., 1999, *Geometric Models for Noncommutative Algebras*, American Mathematical Society, Providence, RI.

Champagnat, N., R. Ferriere, and S. Meleard, 2006, Unifying evolutionary dynamics: From individual stochastic processes to macroscopic models, *Theoretical Population Biology*, 69:297-321.

Cover, T., and H. Thomas, 1991, *Elements of Information Theory*, Wiley, New York.

Dembo, A., and O. Zeitouni, 1998, *Large Deviations and Applications*, 2nd edition, Springer, New York.

Diekmann U., and R. Law, 1996, The dynamical theory of coevolution: a derivation from stochastic ecological processes, *Journal of Mathemaical Biology*, 34:579-612.

Dill, K., S. Banu Ozkan, T. Weikl, J. Chodera, and V. Voelz, 2007, The protein folding problem: when will it be solved? *Current Opinion in Structural Biology*, 17:342-346.

Dobson, C., 2003, Protein folding and misfolding, *Nature*, 426:884-890.

Ellis, R., 1985, *Entropy, Large Deviations, and Statistical Mechanics*, Springer, New York.

Feynman, R., 2000, *Lectures on Computation*, Westview, New York.

Fillit, H., D. Nash, T. Rundek, and A. Zukerman, 2008, *American Journal of Geriatric Pharmacotherapy*, 6:100-118.

Glazebrook, J.F., and R. Wallace, 2009a, Small worlds and red queens in the global workspace: an information-theoretic approach, *Cognitive Systems Research*, 10:333-365,

Glazebrook, J.F., and R. Wallace, 2009b, Rate distortion manifolds as models for cognitive information, *Informatica*, 33:309-345.

Goldschmidt, L., P. Teng, R. Riek, and D. Eisenberg, 2010, Identifying the amylome, proteins capable of forming amyloid-like fibrils, *Proceedings of the National Academy of Sciences*, 107:3487-3492.

Goodsell, D., and A. Olson, 2000, Structural symmetry and protein function, *Annual Reviews of Biophysics and Biomolecular Structure*, 29:105-153.

Gunderson, L., 2000, Ecological resilience in theory and application, *Annual Reviews of Ecological Systematics*, 31:425-439.

Haataja, L., T. Gurlo, C. Huang, and P. Butler, 2008, Islet amylod in type 2 diabetes, and the toxic oligomer hypothesis, *Endocrine Reviews*, 29:303-316.

Holling, C., 1973, Resilience and stability of ecological systems, *Annual Reviews of Ecological Systematics*, 4:1-23.

Khinchin, A., 1957, *Mathematical Foundations of Information Theory*, Dover, New York.

Landau, L., and E. Lifshitz, 2007, *Statistical Physics, Part I*, Elsevier, New York.

Lei, J., and K. Huang, 2010, Protein folding: A perspective from statistical physics.

arXiv:10025013v1.

Levinthal, C., 1968, Are there pathways for protein folding? *Journal de Chimie Physique et de Physicochimie Biologique*, 65:44-45.

Levinthal, C., 1969. In *Mossbauer Spectroscopy*, Debrunner et al. (eds.), University of Illinois Press, Urbana, pp. 22-24.

Onuchic, J., and P. Wolynes, 2004, Theory of protein folding, *Current Opinion in Structural Biology*, 14:70-75.

Pettini, M., 2007, *Geometry and Topology in Hamiltonian Dynamics*, Springer, New York.

Plaxco, K., K. Simons, and D. Baker, 1998, Contact order, transition state placement and the refolding rates of single domain proteins, *Journal of Molecular Biology*, 277:985-994.

Protter, P. 1990, *Stochastic Integration and Differential Equations: A new approach*, Springer, New York.

Qiu, C., M. Kivipelto, and E. von Strauss, 2009, Epidemiology of Alzheimer's disease: occurrence, determinants, and strategies toward intervention, *Dialogues in Clinical Neuroscience*, 11:111-128.

Rockafellar, R., 1970, *Convex Analysis*, Princeton University Press, Princeton, NJ.

Scheuner, D., and R. Kaufman, 2008, The unfolded protein response: a pathway that links insulin demand with β-cell failure and daibetes, *Endocrine Reviews*, 29:317-333.

Sharma, V., V. Kaila, and A. Annila, 2009, Protein folding as an evolutionary process, *Physica A*, 388:851-862.

Tlusty, T., 2007, A model for the emergence of the genetic code as a transition in a noisy information channel, *Journal of Theoretical Biology*, 249:331-342.

Wallace, R., and M. Fullilove, 2008, *Collective Consciousness and its Discontents*, Springer, New York.

Wallace, R., and D. Wallace, 2008, Punctuated equilibrium in statistical models of generalized coevolutionary resilience: how sudden ecosystem transitions can entrain both phenotype expression and Darwinian selection, *Transactions on Computational Systems Biology IX*, LNBI 5121:23-85.

Wallace, R., and D. Wallace, 2009, Code, context, and epigenetic catalysis in gene expression, *Transactions on Computational Systems Biology XI*, LNBI 5750, 283-334.

Wallace, R., and D. Wallace, 2010, Cultural epigenetics: on the heritability of complex diseases. In press, *Transactions on Computational Systems Biology*.

Wallace, R., 2005, *Consciousness: A mathematical treatment of the global neuronal workspace model*, Springer, New York,

Wallace, R., 2009, Metabolic constraints on the eukaryotic transition, *Origins of Life and Evolution of Biospheres*, 39:165-176.

Wallace, R., 2010, Metabolic constraints on the evolution of genetic codes: Did multiple 'preaerobic' ecosystem transitions entrain richer dialects via Serial Endosymbiosis?

http://precedings.nature.com/documents/4120/version/3.

Wallach, J., and M. Rey, 2009, A socioeconomic analysis of obesity and diabetes in New York City, *Public Health Research, Practice, and Policy*, Centers for Disease Control and Prevention,
http://www.cdc.gov/pcd/issues/2009/jul/08$_0$215.*htm*.

Weinstein, A., 1996, Groupoids: unifying internal and external symmetry, *Notices of the American Mathematical Association*, 43:744-752.

Wolynes, P., 1996, Symmetry and the energy landscapes of

biomolecules, *Proceedings of the National Academcy of Sciences*, 93:14249-14255.

Zhang, Q., Y. Wang, and E. Huang, Changes in racial/ethnic disparities in the prevalence of type 2 diabetes by obesity level among US adults, *Ethnicity and Health*, 14:439-457.

Chapter 4

Beyond magic bullets

4.1 Introduction

Western medicine's relentless culturally-determined search for simple magic bullets against complex multifactorial chronic and infectious disease is coming to an end as the low-hanging fruit for such interventions is picked off and as pathogens evolve out from under existing antibiotics. Figure 4.1, adapted from Bernstein, (2010), shows the number of small molecule and biologic USFDA approvals per inflation-adjusted billion dollars in research investment between 1950 and 2010. The cost per intervention has increased from about $ 200 million to over $ 1.2 billion, and many pharmaceutical firms have markedly cut their research efforts.

Paul et al. (2010) summarize the crisis as follows:

> The pharmaceutical industry is facing unprecedented challenges to its business model. Experienced observers and industry analysts have even predicted its imminent demise... Simply stated, without a dramatic increase in [Research and Development] productivity, today's pharmaceutical industry cannot sustain sufficient innovation to replace the loss of revenues due to patent expirations for successful products...
>
> As the pharmaceutical industry transitions from an era of 'me-too' or 'slightly me-better' drugs to one of highly innovative medicines that result in markedly improved health outcomes, it must... re-focus its resources (money and talent) on discovery research and early translational medicine. Although the scientific substrate for drug discovery has never been more abundant, a more complete understanding of human (disease) biology will still be required before many true breakthrough medicines emerge.

129

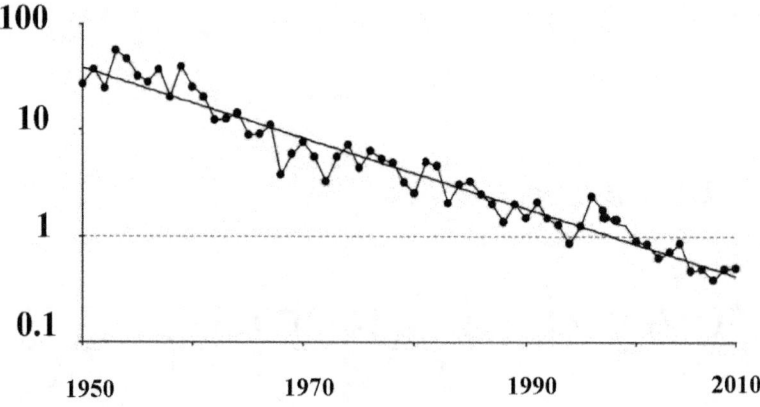

Figure 4.1: Adapted from Bernstein, 2010. The inverse Moore's Law for pharmaceuticals. The number of small molecule and biologic USFDA approvals per inflation-adjusted $ billion in research investment, 1950-2010. The apparent log-linear 'decline in research productivity' represents the failure of complex physiological processes to respond to simple interventions. Western medicine, as defined in the latter half of the 20th Century, has hit a brick wall, a catastrophic regime of exponential cost increase.

Here we will outline a kind of statistical theory of biological interaction that might significantly improve our understanding of human disease biology and contribute to altering the form of figure 4.1, a depressing inverse medical parallel to the famous Moore's Law that has characterized the doubling of on-chip computing power every two years since 1971.

We begin with a review of a canonical formal approach to complicated biological and other dynamics.

4.2   Symbolic dynamics

Symbolic dynamics is a 'coarse-grained' perspective on complicated systems that discretizes their time trajectories in terms of dynamically accessible regions so that it is possible to do statistical mechanics on symbol sequences (e.g., McCauley, 1993, Ch. 8) that can be said to constitute an 'alphabet'. Within that 'alphabet', certain 'statements' are highly probable, and others far less so.

The simple (ideal) oscillating population process described by the equations

$$dX/dt = \omega Y, dY/dt = -\omega X$$

has the solution

$$X(t) = \sin(\omega t), Y(t) = \cos(\omega t)$$

so that

$$X(t)^2 + Y(t)^2 \equiv 1,$$

and the system traces out a circular trajectory in time. Divide the $X-Y$ plane into two components, the simplest possible coarse graining, calling the halfplane to the left of the vertical Y axis A and that to the right B. This system, over units of the period $1/(2\pi\omega)$, traces out a stream of A's and B's having a very precise grammar and syntax:

$$ABABABAB...$$

Many other such statements might be conceivable, e.g.,

$$AAAAAA..., BBBBB..., AAABAAAB..., ABAABAAAB...,$$

and so on, but, of the infinite number of possibilities, only one is actually observed, is 'grammatical'.

More complex dynamical models, incorporating diffusional drift around deterministic solutions, or elaborate structures of complicated stochastic differential equations having various domains of attraction – different sets of 'grammars' – can be described by analogous means (e.g., Beck and Schlogl, 1995, Ch. 3).

Rather than taking symbolic dynamics as a simplification of more exact analytic or stochastic approaches, it is possible to comprehensively generalize the technique itself. Complicated cellular or other physiological processes may not have identifiable sets of stochastic differential equations like noisy, nonlinear mechanical clocks, but, under appropriate coarse-graining, they may still have recognizable grammar and syntax over the long-term. Proper coarse-graining may, however, often be the hard scientific kernel of the problem.

The fundamental assumption for complicated biological developmental phenomena like the onset of infection or the failure of essential regulatory processes is that developmental trajectories can be classified into two groups, a very large set that has essentially zero probability, and a much smaller 'grammatical' set. For the grammatical/syntactical set, the underlying argument is that, given a set of developmental trajectories of length n, the number of grammatical ones, $N(n)$, follows a limit law of the form

$$H = \lim_{n \to \infty} \frac{\log[N(n)]}{n}$$

(4.1)

such that H both exists and is independent of path. If convergence occurs for some finite n_H, then the process is said to be of order n_H. This is a critical foundation of, and limitation on, the modeling strategy adopted here, and constrains its possible realm of applicability. It is, however, fairly general in that it is independent of the serial correlations along reaction pathways.

H is seen to represent the Shannon uncertainty of a classic information source (e.g., Ash, 1990; Cover and Thomas, 2006; Khinchin, 1957).

The basic argument is shown in figure 4.2, where an initial developmental configuration, $\mathbf{S}_0$, can either converge on a normal state $\mathbf{S}_{norm}$ via the set of high probability reaction paths to the left of the filled triangle, or it can converge to a thermodynamically competitive pathological state $\mathbf{S}_{path}$ to the right.

We are, through coarse-graining and symbolic dynamics, assigning classic information sources to the two sets of thermodynamically competitive 'grammatical' pathways. The essential question is how, in general terms, an embedding physiological regulatory structure can act in such a circumstance to change the probabilities of convergence on $\mathbf{S}_{norm}$ or $\mathbf{S}_{path}$.

The mechanism implied by figure 4.2 has, in fact, been observed across much of molecular biology and been a subject of study for decades: Molecular systems may equilibrate between thermodynamically equivalent conformations until one is 'chosen', in a sense, by some external signal. We give some examples, largely taken from the comprehensive review by James and Tawfik (2003):

1. Volkman et al. (2001) show that the protein NtrC is allostatically regulated by phosphorylation: two conformations are present in unphosphorylated NtrC, with phosphorylation merely shifting the equilibrium toward the active conformation.

2. Cordes et al. (2000) show that, with regard to the DNA-binding domain of the Arc repressor, the Arc-N11L mutant spontaneously interconverts between a two-stranded antiparallel β sheet and a two-3_{10}-helix structure. In the absence of a DNA ligand, the two conformations are in equilibrium at almost equal proportions. Addition of the DNA ligand shifts the equilibrium towards the β-sheet form.

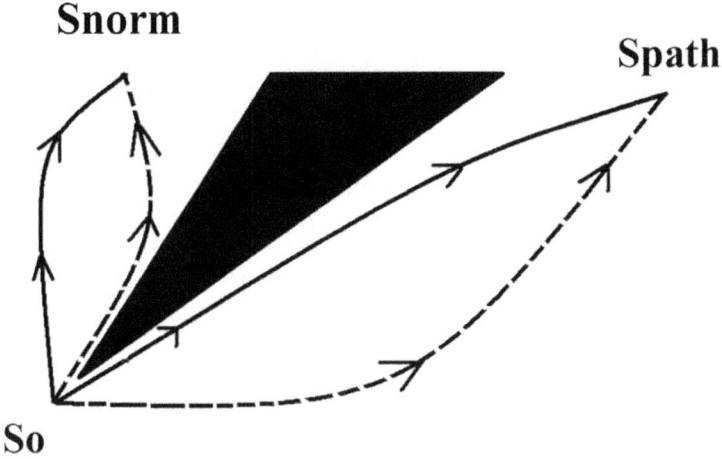

Figure 4.2: An initial physiological configuration S_0 can either develop to a normal final configuration S_{norm} via the set of high probability reaction paths to the left of the filled triangle, or it can converge to a thermodynamically competitive pathological state S_{path} to the right.

3. According to Chen and Weissman (2001), the prion protein PrP interconverts between an α-helix PrP^c and an all β-sheet conformation PrP^{sc}. The β-sheet is trapped by subsequent oligomerization, resulting in amyloid deposit and the onset of disease.

4. Very early on, Pauling (1940) and Landsteiner (1936) proposed that some proteins – antibodies – can exist as an ensemble of isomers with different structures but with similar free energy, so that, if each isomer was able to bind to a different ligand, functional diversity could go far beyond sequence diversity.

5. More recently, Ehrlicher et al. (2011) describe how mechanical strain in actin networks regulates FilGAP and integrin binding to filamin A. When FLNA is mechanically deformed, an otherwise hidden integrin site is exposed, allowing β_7 integrin to bind, while other binding site species are spatially separated, preventing FilGAP from binding. In this case, mechanical deformation provides the external signal.

6. Wallace (2011) has applied an 'information catalysis' model to biological logic gates whose activation or inhibition is triggered by selective binding by an intrinsically disordered protein, and this will serve as the basis for the approach taken here.

4.3   Dual information source

The first step in answering the question of how pathways in figure 4.2 are 'chosen' lies in describing the activity of a large class of regulatory activity in terms of another information source. Atlan and Cohen (1998), in the context of a study of the immune system, argue that the essence of cognition is the comparison of a perceived signal with an internal, learned picture of the world, and then choice of a single response from a large repertoire of possible responses. Such choice inherently involves information and information transmission since it always generates a reduction in uncertainty. Structures that process information are constrained by the asymptotic limit theorems of information theory, in the same sense that sums of stochastic variables are constrained by the Central Limit Theorem, allowing the construction of powerful statistical tools useful for data analysis.

More formally, a pattern of incoming input $\mathbf{S}_i$ describing the status of the physiological system of interest – starting with the initial state $\mathbf{S}_0$ – is mixed in a systematic algorithmic manner with a pattern of otherwise unspecified 'ongoing activity', including cellular, epigenetic and environmental signals, $\mathbf{W}_i$, to create a path of combined signals $x = (a_0, a_1, ..., a_n, ...)$. Each a_k thus represents some functional composition of internal and external factors, and is expressed in terms of the intermediate states as

$$\mathbf{S}_{i+1} = f([\mathbf{S}_i, \mathbf{W}_i]) = f(a_i)$$

(4.2)

for some unspecified function f. The a_i are seen to be very complicated composite objects, in this treatment, that we may choose to coarse-grain so as to obtain an appropriate 'alphabet'.

In a simple spinglass-like model, $\mathbf{S}$ would be a vector, $\mathbf{W}$ a matrix, and f would be a function of their product at 'time' i.

The path x is fed into a highly nonlinear decision oscillator, h, a 'sudden threshold machine' pattern recognition structure, in a sense, that generates an output $h(x)$ that is an element of one of two disjoint sets B_0 and B_1 of possible system responses. Let us define the sets B_k as

$$B_0 = \{b_0, ..., b_k\},$$

$$B_1 = \{b_{k+1}, ..., b_m\}.$$

It is possible to assume an elaborate graded response, in precisely the sense studied by Pufall et al. (2005), supposing that if $h(x) \in B_0$, the pattern is not recognized, and if $h(x) \in B_1$, the pattern has been recognized, and some action $b_j, k + 1 \leq j \leq m$ takes place. Typically, for the example of figure 4.2, the set B_1 would represent the final state of the developing system.

The principal objects of formal interest are paths x triggering pattern recognition-and-response. That is, given a fixed initial state $a_0 = [\mathbf{S}_0, \mathbf{W}_0]$, examine all possible subsequent paths x beginning with a_0 and leading to the event $h(x) \in B_1$. Thus $h(a_0, ..., a_j) \in B_0$ for all $0 < j < m$, but $h(a_0, ..., a_m) \in B_1$. B_1 is thus the set of final possible states, $\{\mathbf{S}_{norm}\} \cup \{\mathbf{S}_{path}\}$ from figure 4.2 that includes both the normal and pathological conditions.

Again, for each positive integer n, let $N(n)$ be the number of high probability grammatical and syntactical paths of length n which begin with some particular a_0 and lead to the condition $h(x) \in B_1$. Call such paths 'meaningful', assuming, not unreasonably, that $N(n)$ will be considerably less than the number of all possible paths of length n leading from a_0 to the condition $h(x) \in B_1$.

While the combining algorithm, the form of the nonlinear oscillator, and the details of grammar and syntax, can all be unspecified in this model, the critical assumption that permits inference of the necessary conditions constrained by the asymptotic limit theorems of information theory is that, again, the finite limit

$$ H = \lim_{n \to \infty} \frac{\log[N(n)]}{n} $$

both exists and is independent of the path x.

Call such a pattern recognition-and-response cognitive process *ergodic*. Not all cognitive processes are likely to be ergodic in this sense, implying that H, if it indeed exists at all, is path dependent, although extension to nearly ergodic processes seems possible (e.g., Wallace and Fullilove, 2008).

Invoking the spirit of the Shannon-McMillan Theorem, as choice involves an inherent reduction in uncertainty, it is then possible to define an adiabatically, piecewise stationary, ergodic (APSE) information source $\mathbf{X}$ associated with stochastic variates X_j having joint and conditional probabilities $P(a_0, ..., a_n)$ and $P(a_n | a_0, ..., a_{n-1})$ such that appropriate conditional and joint Shannon uncertainties satisfy the classic information theory relations (Cover and Thomas, 2006)

$$ H = \lim_{n \to \infty} \frac{\log[N(n)]}{n} = \lim_{n \to \infty} H(X_n | X_0, ..., X_{n-1}) = $$

$$\lim_{n \to \infty} \frac{H(X_0, ..., X_n)}{n+1}.$$

(4.3)

This information source is defined as *dual* to the underlying ergodic cognitive process.

Adiabatic means that the information source has been parameterized according to some scheme, and that, over a certain range, along a particular piece of parameter trajectory, the source remains as close to stationary and ergodic as needed for information theory's central theorems to apply. *Stationary* means that the system's probabilities do not change in time, and *ergodic*, roughly, that the cross sectional means approximate long-time averages. Between pieces it is necessary to invoke various kinds of phase transition formalisms, as described more fully in e.g., Wallace (2005).

4.4   Information catalysis

In the limit of large n, $H = \lim_{n \to \infty} \log[N(n)]/n$ becomes homologous to the free energy density of a physical system at the thermodynamic limit of infinite volume. More explicitly, the free energy density of a physical system having volume V and partition function $Z(\beta)$ derived from the system's Hamiltonian – the energy function – at inverse temperature β is (e.g., Landau and Lifshitz 2007)

$$F[K] = \lim_{V \to \infty} -\frac{1}{\beta} \frac{\log[Z(\beta, V)]}{V} \equiv$$

$$\lim_{V \to \infty} \frac{\log[\hat{Z}(\beta, V)]}{V},$$

(4.4)

with $\hat{Z} = Z^{-1/\beta}$. The latter expression is formally similar to the first part of equation (4.3), a circumstance having deep implications: Feynman (2000) describes in great detail how information and free energy have an inherent duality. Feynman, in fact, defines information precisely as the free energy needed to erase a message. The argument is surprisingly direct (e.g., Bennett, 1988), and for very simple systems it is easy to design a small (idealized) machine that turns the information within a message directly into usable work – free energy. Information is a form of free energy and the construction and transmission of information within living things consumes metabolic free energy, with inevitable losses via the second law of thermodynamics.

Information catalysis, in the circumstance of figure 4.2, arises most simply via the 'information theory chain rule' (Cover and Thomas, 2006). Given X as the information source representing the reaction paths of figure 4.2, and Y, an information source dual to the sophisticated biochemical cognition of the regulating system, one can define jointly typical paths $z_i = (x_i, y_i)$ having the joint information source uncertainty $H(X, Y)$ satisfying

$$H(X, Y) = H(X) + H(Y|X) \leq H(X) + H(Y).$$

(4.5)

Of necessity, then,

$$H(X, Y) < H(X) + H(Y)$$

(4.6)

if $H(Y|X) < H(Y)$.

These relations imply that, by means of the identification of information as a form of free energy, at the expense of adding the considerable energy burden of the regulatory apparatus, represented by its dual information source Y, it becomes possible to canalize the reaction paths of figure 4.2, so as to make one set of pathways beginning with S_0 far more probable than another.

That is, by raising the entire reaction free energy landscape corresponding to $H(X)$ by the amount $H(Y)$ it becomes possible to deepen the energy channel leading from S_0 to the desired outcome, either S_{norm} or S_{path}. Complicated internal reaction mechanisms have been subsumed by the Shannon-McMillan Theorem, in the same sense that the Central Limit Theorem subsumes the behavior of long sums of stochastic variates into the Normal distribution.

Within an organism, however, there will be an ensemble of possible developmental states and pathways, driven by available metabolic free energy, so that, taking $< .. >$ as representing an average,

$$[< H(X,Y) >] < [< H(X) > + < H(Y) >].$$

(4.7)

Typically, letting M represent the intensity of available metabolic free energy, a rate index, one expects

$$< H > \approx \frac{\int H \exp[-H/\kappa M]dH}{\int \exp[-H/\kappa M]dH} \approx \kappa M,$$

(4.8)

where κ, an inverse energy rate scaling constant, may be quite small indeed, a consequence of entropic translation losses between metabolic free energy and the expression of information.

The resulting relation,

$$M_{X,Y} < M_X + M_Y,$$

(4.9)

suggests an explicit free energy mechanism for developmental canalization.

If entropic translation losses are not linear with increase in information transmission rate H, we might replace κM in equation (4.8) with some function $Q(\kappa M)$ that 'tops out' with increasing M, for example $Q \propto \log[\kappa M]$. This means that, after a certain point, large increases in metabolic free energy are needed to increase biological information. The energy relation then becomes, after a little algebra,

$$M_{X,Y} < \kappa \times M_X \times M_Y \ll M_X + M_Y,$$

(4.10)

if either κ or one of the other M-terms is small, and a low energy information source regulator could thus be used to 'leverage' reaction canalization very efficiently.

In reality, there will be a large, nested, set of appropriately coarse-grained regulatory and/or signaling processes expressed as information sources – $Y_1, ..., Y_m$ – in which the physiological system of interest will be embedded, ranging from other physiological systems to patterns of social and cultural interaction, mediated by epigenetic inheritance across generations. Equations (4.7), (4.9) and (4.10) then become

$$[< H(X, Y_1, ..., Y_m) >] < [< H(X) > + \sum_j < H(Y_j) >],$$

$$M_{X,Y_1,...,Y_m} < M_X + \sum_j M_{Y_j},$$

$$M_{X,Y_1,...,Y_m} < \kappa^m M_X \Pi_j M_{Y_j} \ll M_X + \sum_j M_{Y_j}.$$

(4.11)

That is, quite counterintuitively, entropic loss can be a powerful tool for triggering complex biological logic gates like figure 4.2, in much the same sense that Tompa and Csermely (2004) propose that entropy transfer can be used by generalized chaperones to trigger proper conformation in pathologically folded protein complexes.

The last expression in equation (4.11) suggests, in particular, that the appropriate regulatory level for intervention *may not be that of the desired target*. That is, the model implies that the most efficient intervention may be upstream from the desired target or, more likely, involve synergistic dynamic intrusions at more than one scale or level of organization to bring down the overall magnitude of the product term.

We now build a sequence of statistical models based on these foundations.

4.5   No free lunch

The famous 'no free lunch' theorem of Wolpert and Macready (1997) illuminates the next step in the argument. As English (1996) states the matter,

> ...Wolpert and Macready... have established that there exists no generally superior [computational] function optimizer. There is no 'free lunch' in the sense that an optimizer 'pays' for superior performance on some functions with inferior performance on others...
>
> [That is,] superiority on one subset of functions [i.e., problem to be solved, implies] inferiority on the complementary subset...
>
> Hammers contain information about the distribution of nail-driving problems. Screwdrivers contain information about the distribution of screw-driving problems. Swiss army knives contain information about a broad distribution of survival problems. Swiss army knives do many jobs, but none particularly well. When the many jobs must be done under primitive conditions, Swiss army knives are ideal.
>
> The tool literally carries information about the task...

Another way of stating this is to say that a computed solution is simply the product of the information processing of a problem, and, by a very famous argument, information can never be gained simply by processing. Thus a problem X is transmitted as a message by an information processing channel, Y, a computing device, and recoded as an answer. By the 'tuning theorem' argument of the Mathematical Appendix, there will be a channel coding of Y which, when properly tuned, is most efficiently 'transmitted', in a purely formal sense, by the problem. In general, then, the most efficient coding of the transmission channel, that is, the best algorithm turning a problem into a solution, will necessarily be highly problem-specific. Thus there can be no 'best' algorithm for all sets of problems, although there will likely be an optimal algorithm for any given set.

Based on the no free lunch argument, it is clear that different challenges facing a biological entity or subsystem of such an entity must be met by different arrangements of basic biochemical or other cognitive modules. It is possible to make a very abstract picture of this phenomenon, not based on biological network anatomy, but rather on the linkages between the information sources dual to the basic physiological and learned cognitive modules (CM). That is, *the remapped network of interacting CM is reexpressed in terms of the information sources dual to them.* Given two distinct problems classes, there must be two different 'wirings' of the information sources dual to the physiological CM, as in figure 4.3, with the network graph edges measured by the amount of information crosstalk between sets of nodes representing the dual information sources. A more formal treatment of such coupling can be given in terms of network information theory (Cover and Thomas, 2006), particularly incorporating the effects of embedding contexts, implied by the 'external' information source Z – signals from the environment, or simple or composite 'drugs' that may alter the interactions between CM components, and shift the operation of the underlying networks. In the context of this analysis, Z is seen as a therapeutic intervention designed to shift the function of biochemical networks from pathological to benign states.

4.6 Nonequilibrium 'equilibria'

Equation (4.11) allows application of a more sophisticated model that can describe transitions between the two (or more) possible configurations of figure 4.3. The tool for this is a version of Onsager's phenomenological nonequilibrium thermodynamics. Redefining the metabolic free energy intensity measure $M_{X,Y_1,...,Y_m}$ as F, and taking it as parameterized by some set of appropriate variates $\mathbf{Q} = [Q_1, ..., Q_n]$, we can write an 'entropy' in standard form as

$$S \equiv F - \sum_k Q_k \partial F/\partial Q_k.$$

(4.12)

The phenomenological Onsager equation becomes

$$dQ_j/dt = \sum_i L_{i,j} \partial S/\partial Q_i,$$

(4.13)

where the $L_{i,j}$ are empirical constants, and the partial derivatives represent 'thermodynamic forces' driven by gradients in the entropy. It is important to note, however, that, for this system, one cannot have

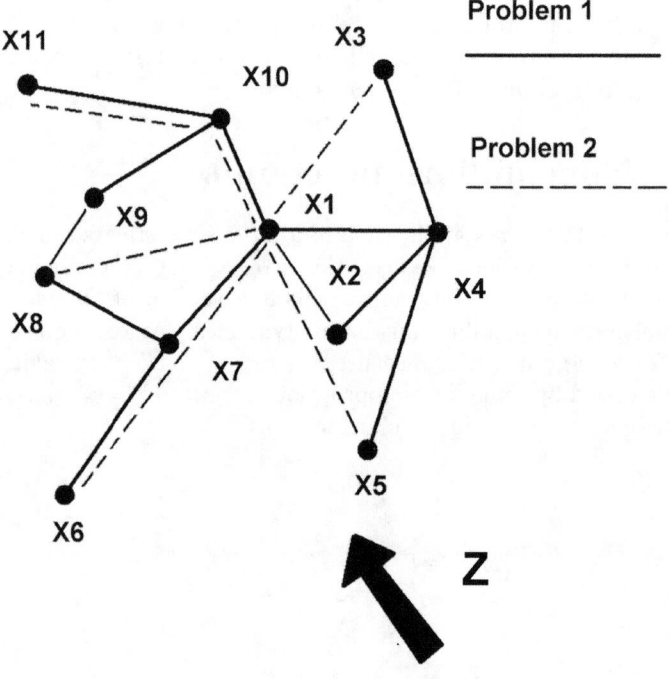

Figure 4.3: By the no free lunch theorem, two markedly different problems will be optimally solved by two different linkages of available biological cognitive modules – characterized now by their dual information sources X_j – into different temporary networks of working structures, here represented by crosstalk among those sources rather than by the exchange of chemicals between physiological CM themselves. The embedding information source Z represents the influence of external signals – environmental perturbations, magic bullet drugs or more comprehensive intervention strategies – whose effects can be accounted for by an application of network information theory.

'reciprocity relations' of the form $L_{i,j} = L_{j,i}$ since the underlying information sources, of which M is an environmental index, are not microreversible. For example, in English the short string ' eht ' does not have the same probability as the equally short string ' the '.

Equation (4.13) has the standard generalization as a stochastic differential equation

$$dQ_t^j = L_j(t, Q^1...Q^n)dt + \sum_j \sigma_j(t, Q^1, ..., Q^n)dB_t^i,$$

(4.14)

where the dB_t^i represent different kinds of 'noise' whose characteristics are usually expressed in terms of their quadratic variation. See, e.g., Zhu et al. (2007) for an example, and any standard work on stochastic differential equations or Brownian motion for a tutorial (e.g., Protter, 1990).

Several points emerge:

1. The different configurations possible to this 'coevolutionary' system that generalizes figure 4.3 are found by setting this set of equations to zero, and solving for stationary points, since the noise terms preclude unstable equilibria.

2. The system may, however, converge to limit cycle or pseudorandom 'strange attractor' behaviors in which it seems to chase its tail endlessly within a limited venue – a kind of 'Red Queen' pathology.

3. What is converged to, in both cases, is not a simple state or limit cycle of states, but rather an equivalence class, or set of them, of highly dynamic information sources coupled by mutual interaction through crosstalk that have simply been parameterized by the free energy intensity measure $F = M_{X,Y_1,...}$. 'Stability' in this structure represents particular patterns of ongoing dynamics rather than some identifiable static configuration.

Most importantly, as Champagnat et al. (2006) note, shifts between the quasi-equilibria of a coevolutionary system like this one can be addressed by the large deviations formalism. They find that the issue of dynamics drifting away from trajectories predicted by the canonical equation can be investigated by considering the asymptotic of the probability of 'rare events' for the sample paths of the diffusion.

By 'rare events' they mean diffusion paths drifting far away from the direct solutions of the canonical equation. The probability of such rare events is governed by a large deviation principle: when a critical parameter (designated ϵ) goes to zero, the probability that the sample path of the diffusion is close to a given rare path ϕ decreases exponentially to 0 with rate $\mathcal{I}(\phi)$, where the 'rate function' $\mathcal{I}$ can be expressed in terms of the parameters of the diffusion. This result, in their view, can be used to study long-time behavior of the diffusion process when there are multiple attractive singularities. Under proper conditions the most likely path followed by the diffusion when exiting a basin of attraction is the one minimizing the rate function $\mathcal{I}$ over all the appropriate trajectories. The time needed to exit the basin is of the order $\exp(V/\epsilon)$ where V is a quasi-potential representing the minimum of the rate function $\mathcal{I}$ over all possible trajectories.

An essential fact of large deviations theory is that the rate function $\mathcal{I}$ which Champagnat et al. invoke can be expressed as a kind of entropy, that is, having the canonical form

$$\mathcal{I} = -\sum_j P_j \log(P_j)$$

(4.15)

for some probability distribution. This result goes under a number of names; Sanov's Theorem, Cramer's Theorem, the Gartner-Ellis Theorem, the Shannon-McMillan Theorem, and so forth (Dembo and Zeitouni, 1998).

These considerations lead very much in the direction of equation (4.14), but now seen as subject to internally-driven large deviations *that are themselves described as information sources*, providing $Q = f(\mathcal{I})$-parameters that can trigger punctuated shifts between quasi-stable modes. Thus both external signals, characterized by the information source Z, and internal 'dynamic ruminations', characterized by the information source $\mathcal{I}$, can provide Q-parameters that serve to drive the system to different quasi-equilibrium states – pathological or benign – in a highly punctuated manner, if they are of sufficient magnitude.

4.7   Discussion and conclusions

Complex multi-level regulatory behaviors, and their failures as affected by environmental interactions or internal dynamics, have been modeled in terms of a nested set of information sources that are constrained by the asymptotic limit theorems of information theory, and this may allow construction of regression- or Onsager- model-like statistical tools useful for scientific inference, focusing on the behaviors of the system rather than on a detailed description of its mechanical state under all circumstances and at all times. The analogy is to characterize the behavior of a computer in terms of its program, rather than attempting provide a full cross-sectional statement of the condition of each logic gate at each clock cycle.

The composite regulatory and/or embedding 'logic gates' affecting figure 4.2 and the generalization of figure 4.3 are likely to be quite different from 'simple' computer models, having extraordinarily subtle properties: evolution is not restricted to binary mathematics (AND, OR, XOR, etc.).

These considerations add considerable weight to an emerging perspective that sees a fundamental defining characteristic of the living state as the operation of chemical or other cognitive processes at virtually all scales and levels of organization (e.g., Wallace, 2011; Wallace and Wallace, 2010; Atlan and Cohen, 1998; Cohen, 2000; Wallace, 2005; Wallace and Fullilove, 2008).

From that viewpoint, the solution to the conundrum of figure 4.1, as expanded by the arguments leading to figure 4.3, is to reconfigure interventions so as to encapsulate more than a single scale or level of organization. That is, it has now become necessary for the pharmaceutical industry to move beyond small molecule design to the principled construction of more comprehensive multifactorial interventions designed to affect the interaction of complementary biochemical and information source networks, driving them from pathological to benign conformations.

At the individual level this would require seeking synergistic total strategies that act across levels of organization, rather than applying a sequence of scale-limited magic bullets, a difficult tectonic shift in scientific perspective, research, and practice not likely to prove popular with those embedded in current funding streams.

At the population level, where public policy can be most effective, the increasing expense of individual level interventions – even if the rate of decline of figure 4.1 can be mitigated as we suggest – would seem to imply the necessity of again recognizing what has been known for the last two hundred years, that patterns of health and illness are determined by living and working conditions and the power relations between groups (e.g., Kleinman, Das and Lock, 1994; Wallace and Fullilove, 2008; Wallace et al., 2009; Wallace and Wallace, 2010).

4.8 References

Ash, R., 1990, *Information Theory*, Dover Publications, NY.

Atlan, H., I. Cohen, 1998, Immune information, self-organization, and meaning, *International Immunology*, 10:711-717.

Beck, C., F. Schlogl, 1995, *Thermodynamics of Chaotic Systems*, Cambridge University Press, NY.

Bennett, C., 1988, Logical depth and physical complexity. In: Herkin, R. (ed.), *The Universal Turing Machine: A Half-Century Survey*, Oxford University Press, pp. 227-257.

Bernstein Research, 2010, *The Long View – R & D Productivity*.

Champagnat, N., R. Ferriere, S. Meleard, 2006, Unifying evolutionary dynamics: from individual stochastic process to macroscopic models, *Theoretical Population Biology*, 69:297-321.

Chien, P., J. Weissman, 2001, Conformational diversity in a yeast prion dictates its seeding specificity, *Nature*, 410: 223-227.

Cohen, I., 2000, *Tending Adam's Garden: Evolving the Cognitive Immune Self*, Academic Press, NY.

Cordes, M., et al., 2000, An evolutionary bridge to a new protein fold, *Nature Structural Biology*, 7:1129-1132.

Cover, T., J. Thomas, 2006, *Elements of Information Theory*, 2nd Edition, Wiley, New York.

Dembo, A., O. Zeitouni, 1998, *Large Deviations and Applications*, Springer, New York.

English, T., 1996, Evaluation of evolutionary and genetic optimizers: no free lunch. In *Evolutionary Programming V: Proceedings of the Fifth Annual Conference on Evolutionary Programming*, Fogel, L. P. Angeline, T. Back (eds.):163-169, MIT Press, Cambridge, MA.

Ehrlicher, A., F. Nakamura, J. Hartwig, D. Weitz, T. Stossel, 2011, Mechanical strain in actin networks regulates FilGAP and integrin binding to filamin A, *Nature*, 478:260-263.

Feynman, R., 2000, *Lectures on Computation*, Westview Press, NY.

Glazebrook, J.F., R. Wallace, 2009, Rate distortion manifolds as model spaces for cognitive information, *Informatica*, 33:309-346.

James, L., D. Tawfik, 2003, Conformational diversity and protein evolution: a 60-year old hypothesis revisited, *Trends in Biochemical Science*, 28:361-368.

Khinchin, A., 1957, *The Mathematical Foundations of Information Theory*, Dover Publications, New York.

Kleinman, A., V. Das, M. Lock, 1994, *Social Suffering*, University of California Press, Berkeley, CA.

Landau, L., E. Lifshitz, 2007, *Statistical Physics, Part I*, Elsevier, NY.

Landsteiner, K., 1936, *The Specificity of Serological Reactions*, Dover Publications, NY, reprinted 1962.

McCauley, J., 1993, Chaos, *Dynamics and Fractals: An algorithmic approach to deterministic chaos*, Cambridge University Press, NY.

Paul, S., D. Mytelka, C. Dunwiddie, C. Persinger, B. Munos, S. Lindborg, A. Schact, 2010, How to improve RD productivity: the pharmaceutical industry's grand challenge, *Nature Reviews: Drug Discovery*, 9:203-214.

Pauling, L., 1940, A theory of the structure and process of formation of antibodies, *Journal of the American Chemical Society*, 62:2643-2657.

Protter, P., 1990, *Stochastic Integration and Differential Equations: A New Approach*, Springer, NY.

Pufall, M., G. Lee, M. Nelson, H. Kang, A Velyvis, L. Kay, L. McIntosh, B. Graves, 2005, *Science*, 309:142-145.

Tompa, P., P. Csermely, 2004, The role of structural disorder in the function of RNA and protein chaperones, *FASEB Journal*, 18:1169-1175.

Tompa, P., C. Szasz, L. Buday, 2005, Structural disorder throws new light on moonlighting, *Trends in Biochemical Sciences*, 30:484-489.

Volkman, B., et al., 2001, Two-state allosteric behavior in a single-domain signaling protein, *Science*, 291:2429-2433.

Wallace, R., 2005, *Consciousness: A Mathematical Treatment of the Global Neuronal Workspace Model*, Springer, NY.

Wallace, R., 2011, Multifunction moonlighting and intrinsically disordered proteins: Information catalysis, nonrigid molecule symmetries, and the 'logic gate' spectrum. In press, *Comptes Rendus Chimie*.

Wallace, R., D. Wallace, R.G. Wallace, 2009, *Farming Human Pathogens: Ecological Resilience and Evolutionary Process*, Springer, NY.

Wallace, R., D. Wallace, 2010, *Gene Expression and its Discontents: The Social Production of Chronic Disease*, Springer, NY.

Wallace, R., M. Fullilove, 2008, *Collective Consciousness and Its Discontents*, Springer, NY.

Wolpert, D., W. Macready, 1997, No free lunch theorems for optimization, *IEEE Transactions on Evolutionary Computation*, 1:67-82.

Zhu, R., A. Rebirio, D. Salahub, S. Kaufmann, 2007, Studying genetic regulatory networks at the molecular level: delayed reaction stochastic models, *Journal of Theoretical Biology*, 246:725-745.

Chapter 5

Summary thoughts

These essays have addressed different aspects of the Western biomedical conundrum from a similar perspective, focusing heavily on 'cognitive' phenomena ranging from cellular protein folding up through nested processes of immune, neural, and social cognition, all embedded in a 'language' of culture powerfully molded by historical trajectory. The tool of choice has been a 'statistics' emerging from the asymptotic limit theorems of information theory, in the same sense that parametric statistics arises from the Central Limit Theorem. The mathematical structure necessary for this is not trivial, invoking groupoid versions of familiar topological structures and arguments. These are more fully explored in Glazebrook and Wallace (2009a, b), and would not be alien to a string theorist.

That such formalism can be so smoothly applied to therapeutic failure, the missing heritability of complex diseases, to protein folding and its disorders, and to the 'inverse Moore's Law' of drug design, while perhaps elegant, is not necessarily comforting. Public health problems, unlike those of medicine, although they may indeed also be expressed as revolving around sets of interacting cognitive phenomena (Wallace and Fullilove, 2007; Wallace et al., 2009, 2010), usually do not require hard-to-design cellular-level 'small molecule' interventions. By and large, clean water, adequate and healthful food, absence of collective and individual threat, respectful treatment at, and substantial control over, occupation – in short, good living and working conditions – have long been recognized as fully adequate for long and healthy life. The Western Great Urban Reform and Labor Movements of the late 19th and early 20th Centuries that involved improvements in political, social, and work conditions, greatly reduced the burden of infectious diseases including cholera, tuberculosis, pneumonia, and so on, long before the invention of pharmaceutical antibiotics. Each of the wars of the 20th century undid the effects of those reforms and unleashed scourges of infectious disease. The story is well known.

Chronic deseases are not immune from these dynamics: cancers and many protein folding disorders, ranging from dementias to diabetes, involve 'lifestyle' factors that are, in fact, little more than individual expressions of embedding social forces (Wallace and Wallace, 2010). Change those social forces (ha!), and the rates of such diseases will plummet. As the United States demonstrates, medical interventions do not suffice to confront the epidemics of obesity-related and viral diseases that are central to the 21st Century.

This is not to say that progress in medicine-as-biological engineering will end anytime soon, at least in the US. The immense and entrenched economic interests of both the pharmaceutical industry and the medical establishment are solidly expressed by the lobbying of government and by a stranglehold over research funding, albeit in the context of the 'decline in pharmaceutical industry productivity' characterized by figure 4.1. What will change, however, is the rate of significant progress. Medical reengineering of the developmental trajectories of basic cellular, organ, and higher level biological and biosocial processes requires a functional grasp of basic principles that will not exist in those circles anytime soon: Think of the arrogance and ignorance in the recently popular term *junk DNA*. While the expression is no longer current, the arrogance and ignorance remain. Indeed, the idea that historically-driven embedding epigenetic and cultural factors can express themselves in 'basic' human biological phenomena of health and illness is still largely alien.

We have demonstrated in these essays that a possible avenue for progress in medicine may exist in a statistics of cognitive phenomena, based on the asymptotic limit theorems of information theory. Developing the theory into a set of routine tools for research will, however, require no small investment, and considerable mathematical and scientific sophistication. Underemployed string theorists and others with similar mathematical skills who have the perverse taste for intractable problems where one can actually do experiments and make observations should take note. There may well be a Higgs boson, and dark matter may be eventually be identified and characterized, but the scientific thicket that constitutes biology, and in particular, its application to the cultural construct of Western biomedicine, is not going to be untangled anytime soon.

References

Glazebrook, J.F., and R. Wallace, 2009a, Small worlds and red queens in the global workspace: an information-theoretic approach, *Cognitive Systems Research*, 10:333-365.

Glazebrook, J.F., and R. Wallace, 2009b, Rate distortion manifolds as models for cognitive information, *Informatica*, 33:309-345.

Wallace, R., and M. Fullilove, 2007, *Collective Consciousness and its Discontents: Institutional distributed cognition, racial policy, and public health in the United States*, Springer, New York.

Wallace, R., R.G. Wallace, and D. Wallace, 2009, *Farming Human Pathogens: Ecological resilience and evolutionary process*, Springer, New York.

Wallace, R., and D. Wallace, 2010, *Gene Expression and its Discontents: The social production of chronic disease*, Springer, New York.

Chapter 6

Mathematical Appendix

6.1 Groupoids

6.1.1 Basic ideas

Following Weinstein (1996) closely, a groupoid, G, is defined by a base set A upon which some mapping – a morphism – can be defined. Note that not all possible pairs of states (a_j, a_k) in the base set A can be connected by such a morphism. Those that can define the groupoid element, a morphism $g = (a_j, a_k)$ having the natural inverse $g^{-1} = (a_k, a_j)$. Given such a pairing, it is possible to define 'natural' end-point maps $\alpha(g) = a_j, \beta(g) = a_k$ from the set of morphisms G into A, and a formally associative product in the groupoid $g_1 g_2$ provided $\alpha(g_1 g_2) = \alpha(g_1), \beta(g_1 g_2) = \beta(g_2)$, and $\beta(g_1) = \alpha(g_2)$. Then the product is defined, and associative, $(g_1 g_2) g_3 = g_1 (g_2 g_3)$.

In addition, there are natural left and right identity elements λ_g, ρ_g such that $\lambda_g g = g = g \rho_g$ (Weinstein, 1996).

An orbit of the groupoid G over A is an equivalence class for the relation $a_j \sim G a_k$ if and only if there is a groupoid element g with $\alpha(g) = a_j$ and $\beta(g) = a_k$. Following Cannas da Silva and Weinstein (1999), we note that a groupoid is called transitive if it has just one orbit. The transitive groupoids are the building blocks of groupoids in that there is a natural decomposition of the base space of a general groupoid into orbits. Over each orbit there is a transitive groupoid, and the disjoint union of these transitive groupoids is the original groupoid. Conversely, the disjoint union of groupoids is itself a groupoid.

The isotropy group of $a \in X$ consists of those g in G with $\alpha(g) = a = \beta(g)$. These groups prove fundamental to classifying groupoids.

If G is any groupoid over A, the map $(\alpha, \beta) : G \to A \times A$ is a morphism from G to the pair groupoid of A. The image of (α, β) is the orbit equivalence relation $\sim G$, and the functional kernel is the union of the isotropy groups. If $f : X \to Y$ is a function, then the kernel of f, $ker(f) = [(x_1, x_2) \in X \times X : f(x_1) = f(x_2)]$ defines an equivalence relation.

Groupoids may have additional structure. As Weinstein (1996) explains, a groupoid G is a topological groupoid over a base space X if G and X are topological spaces and α, β and multiplication are continuous maps. A criticism sometimes applied to groupoid theory is that their classification up to isomorphism is nothing other than the classification of equivalence relations via the orbit equivalence relation and groups via the isotropy groups. The imposition of a compatible topological structure produces a nontrivial interaction between the two structures. Above we have introduced a metric structure on manifolds of related information sources, producing such interaction.

In essence, a groupoid is a category in which all morphisms have an inverse, here defined in terms of connection to a base point by a meaningful path of an information source dual to a cognitive process.

As Weinstein (1996) points out, the morphism (α, β) suggests another way of looking at groupoids. A groupoid over A identifies not only which elements of A are equivalent to one another (isomorphic), but *it also parametizes the different ways (isomorphisms) in which two elements can be equivalent*, i.e., all possible information sources dual to some cognitive process. Given the information theoretic characterization of cognition presented above, this produces a full modular cognitive network in a highly natural manner.

Brown (1987) describes the fundamental structure as follows:

> A groupoid should be thought of as a group with many objects, or with many identities... A groupoid with one object is essentially just a group. So the notion of groupoid is an extension of that of groups. It gives an additional convenience, flexibility and range of applications...
>
> EXAMPLE 1. A disjoint union [of groups] $G = \cup_\lambda G_\lambda, \lambda \in \Lambda$, is a groupoid: the product ab is defined if and only if a, b belong to the same G_λ, and ab is then just the product in the group G_λ. There is an identity 1_λ for each $\lambda \in \Lambda$. The maps α, β coincide and map G_λ to λ, $\lambda \in \Lambda$.
>
> EXAMPLE 2. An equivalence relation R on [a set] X becomes a groupoid with $\alpha, \beta : R \to X$ the two projections, and product $(x, y)(y, z) = (x, z)$ whenever $(x, y), (y, z) \in R$. There is an identity, namely (x, x), for each $x \in X$...

Weinstein (1996) makes the following fundamental point:

> Almost every interesting equivalence relation on a space B arises in a natural way as the orbit equivalence relation of some groupoid G over B. Instead of dealing directly

with the orbit space B/G as an object in the category S_{map} of sets and mappings, one should consider instead the groupoid G itself as an object in the category G_{htp} of groupoids and homotopy classes of morphisms.

The groupoid approach has become quite popular in the study of networks of coupled dynamical systems which can be defined by differential equation models, (e.g., Golubitsky and Stewart 2006).

6.1.2 Global and local symmetry groupoids

Here we follow Weinstein (1996), using his example of a finite tiling.

Consider a tiling of the euclidean plane R^2 by identical 2 by 1 rectangles, specified by the set X (one dimensional) where the grout between tiles is $X = H \cup V$, having $H = R \times Z$ and $V = 2Z \times R$, where R is the set of real numbers and Z the integers. Call each connected component of $R^2 \backslash X$, that is, the complement of the two dimensional real plane intersecting X, a tile.

Let Γ be the group of those rigid motions of R^2 which leave X invariant, i.e., the normal subgroup of translations by elements of the lattice $\Lambda = H \cap V = 2Z \times Z$ (corresponding to corner points of the tiles), together with reflections through each of the points $1/2\Lambda = Z \times 1/2Z$, and across the horizontal and vertical lines through those points. As noted by Weinstein (1996), much is lost in this coarse-graining, in particular the same symmetry group would arise if we replaced X entirely by the lattice Λ of corner points. Γ retains no information about the local structure of the tiled plane. In the case of a real tiling, restricted to the finite set $B = [0, 2m] \times [0, n]$ the symmetry group shrinks drastically: The subgroup leaving $X \cap B$ invariant contains just four elements even though a repetitive pattern is clearly visible. A two-stage groupoid approach recovers the lost structure.

We define the transformation groupoid of the action of Γ on R^2 to be the set

$$G(\Gamma, R^2) = \{(x, \gamma, y | x \in R^2, y \in R^2, \gamma \in \Gamma, x = \gamma y\},$$

with the partially defined binary operation

$$(x, \gamma, y)(y, \nu, z) = (x, \gamma\nu, z).$$

Here $\alpha(x, \gamma, y) = x$, and $\beta(x, \gamma, y) = y$, and the inverses are natural.

We can form the restriction of G to B (or any other subset of R^2) by defining

$$G(\Gamma, R^2)|_B = \{g \in G(\Gamma, R^2) | \alpha(g), \beta(g) \in B\}$$

[1]. An orbit of the groupoid G over B is an equivalence class for the relation

$x \sim_G y$ if and only if there is a groupoid element g with $\alpha(g) = x$ and $\beta(g) = y$.

Two points are in the same orbit if they are similarly placed within their tiles or within the grout pattern.

[2]. The isotropy group of $x \in B$ consists of those g in G with $\alpha(g) = x = \beta(g)$. It is trivial for every point except those in $1/2\Lambda \cap B$, for which it is $Z_2 \times Z_2$, the direct product of integers modulo two with itself.

By contrast, embedding the tiled structure within a larger context permits definition of a much richer structure, i.e., the identification of local symmetries.

We construct a second groupoid as follows. Consider the plane R^2 as being decomposed as the disjoint union of $P_1 = B \cap X$ (the grout), $P_2 = B \backslash P_1$ (the complement of P_1 in B, which is the tiles), and $P_3 = R^2 \backslash B$ (the exterior of the tiled room). Let E be the group of all euclidean motions of the plane, and define the local symmetry groupoid G_{loc} as the set of triples (x, γ, y) in $B \times E \times B$ for which $x = \gamma y$, and for which y has a neighborhood $\mathcal{U}$ in R^2 such that $\gamma(\mathcal{U} \cap P_i) \subseteq P_i$ for $i = 1, 2, 3$. The composition is given by the same formula as for $G(\Gamma, R^2)$.

For this groupoid-in-context there are only a finite number of orbits:

$\mathcal{O}_1$ = interior points of the tiles.

$\mathcal{O}_2$ = interior edges of the tiles.

$\mathcal{O}_3$ = interior crossing points of the grout.

$\mathcal{O}_4$ = exterior boundary edge points of the tile grout.

$\mathcal{O}_5$ = boundary 'T' points.

$\mathcal{O}_6$ = boundary corner points.

The isotropy group structure is, however, now very rich indeed:

The isotropy group of a point in $\mathcal{O}_1$ is now isomorphic to the entire rotation group O_2.

It is $Z_2 \times Z_2$ for $\mathcal{O}_2$.

For $\mathcal{O}_3$ it is the eight-element dihedral group D_4.

For $\mathcal{O}_4, \mathcal{O}_5$ and $\mathcal{O}_6$ it is simply Z_2.

These are the 'local symmetries' of the tile-in-context.

6.2   Martingales and SDE's

6.2.1   Martingales

Suppose we have entered one of the great gambling casinos of the world, host to an almost infinite variety of games of chance: card games ranging from baccarat and blackjack to keno and poker, roulette wheels, one armed-bandits, dice games, and so on. Each game has different rules of play, even if, as for card games, the instruments of play are all the same. Complicated outcomes for those instruments produce equally complex patterns of loss or gain for the player.

Suppose a player begins with an initial fortune of some given amount, and bets $n = 1, 2, ...$ times according to a stochastic process in which a stochastic variable $\mathbf{X}_n$, which represents the size of the player's fortune at play n, takes values $\mathbf{X}_n = x_{n,i}$ with probabilities $P_{n,i}$ such that $\sum_i P_{n,i} = 1$, where i represents a particular outcome at step n.

Assume for all n there exists a value $0 < C < \infty$ such that the expectation of $\mathbf{X}_n$,

$$E(\mathbf{X}_n) \equiv \sum_i x_{n,i} P_{n,i} < C$$

(6.1)

for all n. That is, no infinite or endlessly increasing fortunes are permitted.

We note that the state $\mathbf{X}_n = 0$, having probability P_n^0, i.e. the loss of all a player's funds, terminates the game.

We suppose it possible to define conditional probabilities at step $n + 1$ which depend on the way in which the value of $\mathbf{X}_n$ was reached, so that we can define the conditional expectation of $\mathbf{X}_{n+1}$:

$$E(\mathbf{X}_{n+1}|\mathbf{X}_1, \mathbf{X}_2, ...\mathbf{X}_n) \equiv E(\mathbf{X}_{n+1}|n)$$

The 'sample space' for the probabilities defining this conditional expectation is the set of different possible sequences of the $x_{m,i} > 0$: $x_{1,i}, x_{2,j}, x_{3,k}...x_{n,q}$

We call the sequence of stochastic variables $\mathbf{X}_n$ defining the game a *Submartingale* if, at each step n,

$$E(\mathbf{X}_{n+1}|n) \geq \mathbf{X}_n,$$

a *Martingale* if

$$E(\mathbf{X}_{n+1}|n) = \mathbf{X}_n$$

and a *Supermartingale* if

$$E(\mathbf{X}_{n+1}|n) \leq \mathbf{X}_n.$$

$\mathbf{X}_n$ is, remember, the player's fortune at step n.

Clearly a submartingale is favorable to the player, a martingale is an absolutely fair game, and a supermartingale is favorable to the house.

Regardless of the complexity of the game, the details of the playing instruments, the ways of determining gains or loss or their amounts, or any other structural factors of the underlying stochastic process, the essential content of the Martingale Limit Theorem is that in all three cases the sequence of stochastic variables $\mathbf{X}_n$ converges in probability 'almost everywhere' to a well-defined stochastic variable $\mathbf{X}$ as $n \to \infty$. That is, for each kind of martingale, no matter the actual sequence of winnings $x_{1,i}, x_{2,j}, ... x_{n,k}, x_{n+1,m}, ...$, you get to the same limiting stochastic variable $\mathbf{X}$. Sequences for which this does not happen have zero probability.

A simple proof of this result (Petersen, 1995) runs to several pages of dense mathematics using modern theories of abstract integration on sets. Indeed, all the asymptotic theorems we have cited require more or less arduous application of measure theory and Lebesgue integration, topics which are themselves relatively straightforward, elegant and worth study (Rudin, 1976, Royden, 1968). Proofs using more elementary approaches (Karlin and Taylor, 1975, Ch. 6) run to full chapters.

6.2.2   Nested Martingales

We are interested in a compound stochastic process in which the 'winnings' at the 'smaller' scale, played by one set of rules, contribute, in some sense, to a quite different game having completely different rules on a 'larger' scale. These games are bounded by the condition $E(\mathbf{X}_n) < C$, for some finite positive C.

The essential point is that a proportion of the winnings from the smaller game are duplicated by a 'benefactor' and directly raise the magnitude of the player's fortune for the larger, embedding game.

If the inner game is characterized at step n by the random variable $\mathbf{Y}_n$, then the 'real' winnings at step $n + 1$ for the embedding game, associated with the random variable $\mathbf{X}_{n+1}$, become, for some function f_n, which may involve additional stochastic variables,

$$\mathbf{X}_{n+1} = f_n(\mathbf{X}_n, \mathbf{Y}_n, \mathbf{Y}_{n+1}).$$

(6.2)

A slightly different approach would involve conditional expectations in the convolution of scales:

$$E(\mathbf{X}_{n+1}|n) = F_n(\mathbf{X}_n, \mathbf{Y}_n, E(\mathbf{Y}_{n+1}|n))$$

(6.3)

for some function F_n.

Traditionally the simplest version of this extension assumes that the compound game is, in some sense, a subset of the original:

$$\mathbf{X}_{n+1} = \mathbf{X}_n + \mathbf{A}_n(\mathbf{Y}_{n+1} - \mathbf{Y}_n).$$

(6.4)

We assume the filter $\mathbf{A}_n \geq 0$ is a non-negative stochastic variable, which can indeed take the value 0. This may, for example, be greater than zero only one time in ten or a hundred, on average. Taking the conditional expectation gives

$$E(\mathbf{X}_{n+1}|n) = \mathbf{X}_n + \mathbf{A}_n(E(\mathbf{Y}_{n+1}|n) - \mathbf{Y}_n)$$

(6.5)

where we recognize the conditional expectation of any variate $\mathbf{Z}_n$ at step n is just its value.

Since $\mathbf{A}_n \geq 0$, *the game described by the attenuated sequence* $\mathbf{X}_n$ *has the same martingale classification as does the nested central city game described by* $\mathbf{Y}_n$.

6.2.3   The Martingale Transform

The X-processes in equation (6.4) is the *Martingale transform* of $\mathbf{Y}_n$ (Taylor, 1996, p.232; Billingsley, 1968, p. 412), and the result is classic, representing the *impossibility of a successful betting system*.

Note that the basic Martingale transform can be rewritten as

$$\frac{\mathbf{X}_{n+1} - \mathbf{X}_n}{\mathbf{Y}_{n+1} - \mathbf{Y}_n} \equiv \frac{\Delta\mathbf{X}_n}{\Delta\mathbf{Y}_n} = \mathbf{A}_n,$$

(6.6)

or

$$\Delta\mathbf{X}_n = \mathbf{A}_n\Delta\mathbf{Y}_n.$$

(6.7)

Induction gives

$$\mathbf{X}_{n+1} = \mathbf{X}_0 + \sum_{j=1}^{n} \mathbf{A}_j\Delta\mathbf{Y}_j.$$

(6.8)

This notation is suggestive: in fact the Martingale transform is the discrete analog of Ito's stochastic integral relative to a sequence of stopping times, (Taylor, 1996, p. 232; Protter, 1990, p. 44; Ikeda and Watanabe, 1989, p. 48). In the stochastic integral context the Y-process is called the 'integrator' and the A-process the 'integrand.' Further development leads toward generalizations of Brownian motion, the Poisson process, and so on (Meyer, 1989; Protter, 1990).

The basic picture is of the transmission of a signal, $\mathbf{Y}_n$, in the presence of noise, $\mathbf{A}_n$.

6.2.4 Stochastic Differential Equations

A more realistic extension of the elementary denumerable Martingale transform for our purposes is

$$\mathbf{X}_{n+1} = \mathbf{X}_n + (\mathbf{B}_{n+1} - \mathbf{B}_n)\mathbf{X}_n + \mathbf{A}_n(\mathbf{Y}_{n+1} - \mathbf{Y}_n),$$

(6.9)

where $\mathbf{B}_n$ is another stochastic variable.

Using the more suggestive notation of equations (6.6) and (6.7) this becomes the fundamental stochastic differential equation

$$\Delta\mathbf{X}_n = \mathbf{X}_n\Delta\mathbf{B}_n + \mathbf{A}_n\Delta\mathbf{Y}_n.$$

(6.10)

Taking conditional expectations gives

$$E(\mathbf{X}_{n+1}|n) - \mathbf{X}_n =$$

$$\mathbf{X}_n(E(\mathbf{B}_{n+1}|n) - \mathbf{B}_n) + \mathbf{A}_n(E(\mathbf{Y}_{n+1}|n) - \mathbf{Y}_n).$$

(6.11)

If $\mathbf{X}_n, \mathbf{A}_n \geq 0$, the martingale classification of $\mathbf{X}$ depends on those of $\mathbf{B}$ and $\mathbf{Y}$.

Extending the argument to a hierarchically-linked network is straightforward, leading to the Ito stochastic integral

$$\mathbf{X}_{n+1} \approx \mathbf{X}_0 + \sum_{k=1}^{n} \mathbf{A}_k\Delta\mathbf{Y}_k.$$

(6.12)

The complete hierarchical system, then undergoes an iterative Z-process defined by the integrator $\mathbf{X}_j$:

$$\mathbf{Z}_{m+1} \approx \mathbf{Z}_0 + \sum_{j=1}^{m} \mathbf{C}_j \Delta \mathbf{X}_j.$$

(6.13)

Extension of this development to intermediate times is complicated and involves taking the continuous limit of the Riemann-type sums of equations (6.7), (6.11) and (6.12). This produces the stochastic differential equation

$$d\mathbf{X}_t = \mathbf{X}_t d\mathbf{B}_t + \mathbf{A}_t d\mathbf{Y}_t$$

(6.14)

whose solution depends critically on the behavior of the second-order step-by-step 'quadratic variation,' a variance-like limit of the stochastic processes. Letting $\mathbf{U}_n, \mathbf{V}_n$ be two arbitrary processes with $\mathbf{U}_0 = \mathbf{V}_0 = 0$, their quadratic variation is

$$[\mathbf{U}_n, \mathbf{V}_n] \equiv \sum_{j=1}^{n-1} (\mathbf{U}_{j+1} - \mathbf{U}_j)(\mathbf{V}_{j+1} - \mathbf{V}_j).$$

(6.15)

Taking the 'infinitesimal limit' of continuous time, a term-by-term expansion of this sum can be shown to give (e.g., Meyer, 1989; Protter, 1990)

$$[\mathbf{U}_t, \mathbf{V}_t] = \mathbf{U}_t \mathbf{V}_t - \int_0^t \mathbf{U}_s d\mathbf{V}_s - \int_0^t \mathbf{V}_r d\mathbf{U}_r.$$

(6.16)

To put this in some perspective, classical Brownian motion has the 'structure equation' $[\mathbf{X}_t, \mathbf{X}_t] = t$.

That is, for Brownian motion the jump-by-jump quadratic variation increases linearly with time. While much of the contemporary theory of financial markets is based on Brownian analogs, real processes are likely to be more complex, subject to sudden, massive, discontinuous 'phase changes' which cannot be simply characterized as diffusional.

The solution of equation (6.13) is a classic result in the theory of stochastic differential equations (Protter, 1990). We assume for simplicity no discontinuous jumps, and first study the 'exponential' equation

$$d\mathbf{X}_t = \mathbf{X}_t d\mathbf{B}_t$$

or equivalently

$$\mathbf{X}_t = \mathbf{X}_0 + \int_0^t \mathbf{X}_s d\mathbf{B}_s$$

(6.17)

Following Protter (1990, p. 78) this has the solution

$$\mathbf{X}_t = \epsilon(\mathbf{B})_t = \mathbf{X}_0 \exp(\mathbf{B}_t - 1/2[\mathbf{B}_t, \mathbf{B}_t]).$$

(6.18)

Next we define

$$\mathbf{H}_t \equiv \int_0^t \mathbf{A}_s d\mathbf{Y}_s.$$

(6.19)

Equation (6.13) can be restated as

$$\mathbf{X}_t = \mathbf{H}_t + \mathbf{X}_0 + \int_0^t \mathbf{X}_s d\mathbf{B}_s.$$

(6.20)

For the continuous case, this has the formal solution (Protter, 1990, p.266)

$$\epsilon_{\mathbf{H}}(\mathbf{B})_t =$$

$$\epsilon(\mathbf{B})_t [\mathbf{H}_0 + \int_0^t 1/\epsilon(\mathbf{B})_s d(\mathbf{H}_s - [\mathbf{H}, \mathbf{B}]_s)],$$

(6.21)

with

$$1/\epsilon(\mathbf{B}) = \epsilon(-\mathbf{B} + [\mathbf{B}, \mathbf{B}]).$$

(6.22)

The structure equations defining $[\mathbf{B}, \mathbf{B}]$ and $[\mathbf{H}, \mathbf{B}]$ are critical in determining transient behavior, but not likely to have simple Brownian form.

6.3 The 'Tuning Theorem'

Messages from an information source, seen as symbols x_j from some alphabet, each having probabilities P_j associated with a random variable X, are 'encoded' into the language of a 'transmission channel', a random variable Y with symbols y_k, having probabilities P_k, possibly with error. Someone receiving the symbol y_k then retranslates it (without error) into some x_k, which may or may not be the same as the x_j that was sent.

More formally, the message sent along the channel is characterized by a random variable X having the distribution

$$P(X = x_j) = P_j, j = 1, ..., M.$$

The channel through which the message is sent is characterized by a second random variable Y having the distribution

$$P(Y = y_k) = P_k, k = 1, ..., L.$$

Let the joint probability distribution of X and Y be defined as

$$P(X = x_j, Y = y_k) = P(x_j, y_k) = P_{j,k}$$

and the conditional probability of Y given X as

$$P(Y = y_k | X = x_j) = P(y_k | x_j).$$

Then the Shannon uncertainty of X and Y independently and the joint uncertainty of X and Y together are defined respectively as

$$H(X) = -\sum_{j=1}^{M} P_j \log(P_j)$$

$$H(Y) = -\sum_{k=1}^{L} P_k \log(P_k)$$

$$H(X, Y) = -\sum_{j=1}^{M} \sum_{k=1}^{L} P_{j,k} \log(P_{j,k}).$$

The *conditional uncertainty* of Y given X is defined as

$$H(Y|X) = -\sum_{j=1}^{M} \sum_{k=1}^{L} P_{j,k} \log[P(y_k | x_j)]$$

For any two stochastic variates X and Y, $H(Y) \geq H(Y|X)$, as knowledge of X generally gives some knowledge of Y. Equality occurs only in the case of stochastic independence.

Since $P(x_j, y_k) = P(x_j)P(y_k|x_j)$, we have $H(X|Y) = H(X,Y) - H(Y)$.

The information transmitted by translating the variable X into the channel transmission variable Y – possibly with error – and then re-translating without error the transmitted Y back into X is defined as

$$I(X|Y) \equiv H(X) - H(X|Y) = H(X) + H(Y) - H(X,Y)$$

Again, see Ash (1990), Cover and Thomas (2006) or Khinchin (1957) for details. The essential point is that if there is no uncertainty in X given the channel Y, then there is no loss of information through transmission. In general this will not be true, and herein lies the essence of the theory.

Given a fixed vocabulary for the transmitted variable X, and a fixed vocabulary and probability distribution for the channel Y, we may vary the probability distribution of X in such a way as to maximize the information sent. The capacity of the channel is defined as

$$C \equiv \max_{P(X)} I(X|Y)$$

subject to the subsidiary condition that $\sum P(X) = 1$.

The critical trick of the Shannon Coding Theorem for sending a message with arbitrarily small error along the channel Y at any rate $R < C$ is to encode it in longer and longer 'typical' sequences of the variable X; that is, those sequences whose distribution of symbols approximates the probability distribution $P(X)$ above which maximizes C.

If $S(n)$ is the number of such 'typical' sequences of length n, then

$$\log[S(n)] \approx nH(X)$$

where $H(X)$ is the uncertainty of the stochastic variable defined above. Some consideration shows that $S(n)$ is much less than the total number of possible messages of length n. Thus, as $n \to \infty$, only a vanishingly small fraction of all possible messages is meaningful in this sense. This observation, after some considerable development, is what allows the Coding Theorem to work so well. In sum, the prescription is to encode messages in typical sequences, which are sent at very nearly the capacity of the channel. As the encoded messages become longer and longer, their maximum possible rate of transmission without error approaches channel capacity as a limit. Again, the standard references provide details.

This approach can be, in a sense, inverted to give a 'tuning theorem' variant of the coding theorem.

Telephone lines, optical wave guides and the tenuous plasma through which a planetary probe transmits data to earth may all be viewed in traditional information-theoretic terms as a *noisy channel* around which we must structure a message so as to attain an optimal error-free transmission rate.

Telephone lines, wave guides and interplanetary plasmas are, relatively speaking, fixed on the timescale of most messages, as are most sociogeographic networks. Indeed, the capacity of a channel, is defined by varying the probability distribution of the 'message' process X so as to maximize $I(X|Y)$.

Suppose there is some message X so critical that its probability distribution must remain fixed. The trick is to fix the distribution $P(x)$ but *modify the channel* – i.e., tune it – so as to maximize $I(X|Y)$. The *dual* channel capacity C^* can be defined as

$$C^* \equiv \max_{P(Y),P(Y|X)} I(X|Y).$$

But

$$C^* = \max_{P(Y),P(Y|X)} I(Y|X)$$

since

$$I(X|Y) = H(X) + H(Y) - H(X,Y) = I(Y|X).$$

Thus, in a purely formal mathematical sense, *the message transmits the channel*, and there will indeed be, according to the Coding Theorem, a channel distribution $P(Y)$ which maximizes C^*.

One may do better than this, however, by modifying the channel matrix $P(Y|X)$. Since

$$P(y_j) = \sum_{i=1}^{M} P(x_i)P(y_j|x_i),$$

$P(Y)$ is entirely defined by the channel matrix $P(Y|X)$ for fixed $P(X)$ and

$$C^* = \max_{P(Y),P(Y|X)} I(Y|X) = \max_{P(Y|X)} I(Y|X).$$

Calculating C^* requires maximizing the complicated expression

$$I(X|Y) = H(X) + H(Y) - H(X,Y)$$

which contains products of terms and their logs, subject to constraints that the sums of probabilities are 1 and each probability is itself between 0 and 1. Maximization is done by varying the channel matrix terms $P(y_j|x_i)$ within the constraints. This is a difficult problem in nonlinear optimization. However, for the special case $M = L$, C^* may be found by inspection:

If $M = L$, then choose

$$P(y_j|x_i) = \delta_{j,i}$$

where $\delta_{i,j}$ is 1 if $i = j$ and 0 otherwise. For this special case

$$C^* \equiv H(X)$$

with $P(y_k) = P(x_k)$ for all k. *Information is thus transmitted without error when the channel becomes 'typical' with respect to the fixed message distribution $P(X)$.*

If $M < L$ matters reduce to this case, but for $L < M$ information must be lost, leading to Rate Distortion limitations.

Thus modifying the channel may be a far more efficient means of ensuring transmission of an important message than encoding that message in a 'natural' language which maximizes the rate of transmission of information on a fixed channel.

We have examined the two limits in which either the distributions of $P(Y)$ or of $P(X)$ are kept fixed. The first provides the usual Shannon Coding Theorem, and the second a tuning theorem variant, a tunable, retina-like, Rate Distortion Manifold, in the sense of Glazebrook and Wallace (2009).

6.4   References

Ash, 1990, *Information Theory*, Dover Publications, NY.

Billingsley, P., 1968, *Convergence of Probability Measures*, John Wiley and Sons, New York.

Brown, R., 1987, From groups to groupoids: a brief survey, *Bulletin of the London Mathematical Society*, 19:113-134.

Cannas da Silva, A., and A. Weinstein, 1999, *Geometric Models for Noncommutative Algebras*, American Mathematical Society, New Yoprk.

Cover, T., J. Thomas, 2006, *Elements of Information Theory*, 2nd Ed., Wiley, New York.

Glazebrook, J.F., R. Wallace, 2009, Rate distortion manifolds as model spaces for cognitive systems, *Informatica*, 33:309-346.

Golubitsky, M., and I. Stewart, 2006, Nonlinear dynamics and networks: the groupoid formalism, *Bulletin of the American Mathematical Society*, 43:305-364.

Ikeda, N., and S. Watanabe, 1989, *Stochastic Differential Equations and Diffusion Processes*, 2nd edition, North Holland Publishing Co., Amsterdam.

Karlin, S., and H.Taylor, 1975, *A First Course in Stochastic Processes*, 2nd edition, Academic Press, New York.

Khinchin, A., 1957, *The Mathematical Foundations of Information Theory*, Dover Publications, New York.

Meyer, P., 1989, A short presentation on stochastic calculus. Appendix to *Stochastic Calculus on Manifolds*, M. Emery, Springer, New York.

Petersen, K., 1995, *Ergodic Theory*, Cambridge Studies in Advanced Mathematics 2, Cambridge University Press, Cambridge, UK.

Protter, P., 1990, *Stochastic Integration and Differential Equations: A New Approach*, Springer, New York.

Royden, H., 1968, Real Analysis, Macmillan, New York.

Rudin, W., 1976, *Principles of Mathematical Analysis*, McGraw-Hill, New York.

Weinstein, A., 1996, Groupoids: unifying internal and external symmetry, *Notices of the American Mathematical Society*, 43:744-752.

Index

www.ingramcontent.com/pod-product-compliance
Lightning Source LLC
Chambersburg PA
CBHW051516170526
45165CB00002B/487